Viktor Müller

Micromechanical modeling of short-fiber reinforced composites

Schriftenreihe
Kontinuumsmechanik im Maschinenbau
Band 6

Karlsruher Institut für Technologie (KIT)
Institut für Technische Mechanik
Bereich Kontinuumsmechanik

Hrsg. Prof. Dr.-Ing. habil. Thomas Böhlke

Eine Übersicht aller bisher in dieser Schriftenreihe erschienenen Bände finden Sie am Ende des Buchs.

Micromechanical modeling of short-fiber reinforced composites

by
Viktor Müller

KIT Scientific Publishing

Dissertation, Karlsruher Institut für Technologie (KIT)
Fakultät für Maschinenbau
Tag der mündlichen Prüfung: 01. Juli 2015

Impressum

KIT Scientific Publishing

Karlsruher Institut für Technologie (KIT)
KIT Scientific Publishing
Straße am Forum 2
D-76131 Karlsruhe

KIT Scientific Publishing is a registered trademark of Karlsruhe Institute of Technology. Reprint using the book cover is not allowed.

www.ksp.kit.edu

Print on Demand 2016

ISSN 2192-693X
ISBN 978-3-7315-0454-2
DOI 10.5445/KSP/1000050760

Micromechanical modeling of short-fiber reinforced composites

Zur Erlangung des akademischen Grades
Doktors der Ingenieurwissenschaften

bei der Fakultät für Maschinenbau
des Karlsruher Instituts für Technologie (KIT)

genehmigte

Dissertation

von

Dipl.-Ing. Viktor Müller

Datum der mündlichen Prüfung:	01. Juli 2015
Referent:	Prof. Dr.-Ing. habil. Thomas Böhlke
Korreferent:	Prof. Dr.-Ing. habil. Stefan Kolling

Summary

The mechanical behavior of a short-fiber reinforce composite is significantly governed by its microstructure. The microstructure of short-fiber reinforced composites shows heterogeneities on different length scales concerning micro-structural properties like the fiber volume fraction and the fiber orientation distribution.

This work is focused on the prediction of the elastic behavior of short-fiber reinforced composites. For this purpose, a self-consistent homogenization method, the interaction direct derivative estimate, and a two-step bounding method are considered. These mean-field approaches account for detailed microstructure data experimentally determined by micro-computed tomography and, additionally, virtually generated microstructure data.

Firstly, the predictions of the elastic behavior of the homogenization methods are compared with experimental measurements. Secondly, these mean-field methods are contrasted with a full-field voxel-based homogenization approach. Thirdly, based on the class of materials with transversally isotropic fiber orientation distributions, it is investigated, whether the second-order orientation tensor delivers a sufficient microstructure description for the prediction of the elastic properties of the composite.

Zusammenfassung

Das mechanische Verhalten kurzfaserverstärkter Werkstoffe wird maßgeblich durch ihre Mikrostruktur bestimmt. Diese Mikrostruktur weist Heterogenitäten bezüglich des Volumenanteils und der Orientierungsverteilung der Fasern auf verschiedenen Längenskalen auf.

Diese Arbeit beschäftigt sich mit der Vorhersage des elastischen Verhaltens von kurzfaserverstärkten Werkstoffen. Hierfür werden die Selbstkonsistenzmethode, die Interaction-Direct-Derivative Schätzung und ein Zwei-Schritt Ansatz betrachtet. Diese Mean-Field-Ansätze ermöglichen es, detaillierte Mikrostrukturinformationen zu berücksichtigen, die entweder mitttels Mikro-Computer-Tomographie experimentell ermittelt oder künstlich generiert wurden.

Zunächst werden die Vorhersagen des elastischen Verhaltens der Homogenisierungsmethoden mit experimentellen Messungen verglichen. Dann werden die Mean-Field-Ansätze einem voxelbasierten numerischen Homogenisierungsverfahren gegenübergestellt. Zuletzt wird basierend auf der Klasse von Materialien mit transversalisotroper Faserverteilung diskutiert, ob der Orientierungstensor zweiter Stufe ausreichend für die Berechnung elastischer Eigenschaften ist.

Acknowledgment

In the first place, I thank Prof. Thomas Böhlke for the possibility to work at ITM, for the supervision of my doctoral thesis and for his support and confidence. In the past four years, it was a pleasure for me to learn from him a lot concerning professional and non-professional issues.
I thank Prof. Stefan Kolling for his engagement as second reviewer. Thank you for the straightforward communication and the fruitful collaboration.
I very much appreciate the collaborations with Matthias Kabel and Heiko Andrä from Fraunhofer ITWM and Felix Dillenberger and Robert Glöckner from Fraunhofer LBF. Many thanks to all of you for your contributions to our collaboration.
It would not be the same impressive experience without my friends at ITM. Thank you for the professional, personal and humorous discussions during coffee breaks, lunches, conferences and private meetings. Rudolf, I will never forget the supportive atmosphere in our office! Many thanks to Helga and Ute for the proof reading of our papers and for your support in administrative issues. Thank you, Tom, for your help in technical questions and for the helpful discussions on the daily way home by train.
A very special thank goes to my wife Anna. I am so happy to enjoy your presence every day.

Contents

1 Introduction

1.1 Motivation

Nowadays, fiber reinforced composites are increasingly used for different kinds of application. These composites play a very important role in, e.g., the aerospace, the automotive and the sport equipment industry. The main driving force for this trend is the growing effort to obtain lightweight constructions and, thus, to increase the performance, the efficiency and the ease of use of the products. Simultaneously, composite materials have to be relatively cheap, comparatively easy to process and, also, they have to provide advantageous specific mechanical properties in order to justify their substantial market share.

Even though, fiber reinforced composites are used for more and more applications, a robust dimensioning of light-weight constructions with reinforced materials is still a challenging task. This is due to the fact, that fiber reinforced composites show heterogeneities on different length scales concerning microstructural properties like fiber volume fraction and fiber orientation distribution (Fu and Lauke, 1996; Fu et al., 2000). In addition, the microstructural properties depend on the particular manufacturing process and the geometry of the target part.

The essential element of the microstructure of fiber reinforced composites is the fiber itself. The geometry of the fibers is often used to classify these materials into composites with continuous and discontinuous fibers. A typical representative of the first-mentioned, referred to as continuous fiber reinforce composites (CoFRC), are materials with unidirectionally aligned woven fibers. Short-fiber reinforced composites (SFRCs) are typical representatives of the discontinuous fiber reinforced composites (DiCoFRC).

The focus of the present work lies on SFRCs consisting of a thermoplastic matrix and glass fibers. These materials are, usually, manufactured by injection molding. The microstructure of SFRCs is characterized by the spatial distribution, the orientation distribution, the distribution of the aspect-ratio, and the volume fraction of the fibers. Additionally, the microstructure is affected by the manufacturing process. Particularly in shell-like injection-molded parts, the fibers are oriented in layers. In the boundary layer, the fibers are mostly oriented in the filling direction, and in the core layer in the transversal direction, which is perpendicular to the filling direction. The melted polymer behaves like a viscous fluid and the orientation distribution of the fibers can be attributed to the complex flow conditions in the mold cavity with a variation of the flow velocity perpendicular to the flow direction. In addition to this velocity gradient, the orientation distribution of the fibers is determined by the geometrical properties of the fibers itself, the properties of the matrix material, the manufacturing process parameters and the geometry of the mold cavity. The thickness of the core layer increases with increasing viscosity of the polymer melt, with increasing fiber volume fraction and aspect ratio of the fibers and with increasing injection speed of the material (Horst, 1997).

In consequence of the complexity of the microstructure of SFRCs, a reliable prediction of the mechanical properties depends on the particularities of the microstructure itself. In contrary to phenomenological approaches, micromechanically based methods allow for a separate modeling of the properties of each constituent and the interaction between them. Additionally, considering the relation of the microstructure characteristics, the manufacturing process and the effective properties, conclusions on the manufacturing process can be made. Thus, micromechanical approaches permit a deeper understanding of the material behavior, which may be used to increase the performance of structural parts made of such composites.

Thus, a reliable prediction of the mechanical properties depends on the particularities of the microstructure. On the one hand, this complex microstructure can be analyzed experimentally in order to obtain the corresponding descriptors. On the other hand, numerical simulations of the manufacturing process as, e.g., injection molding simulations, can be applied to predict the microstructure configuration. Considering the huge amount of scientific work, which has been accomplished in this field (e.g., Milton (2002) and Dvorak (2012)), the aim to incorporate the microstructure of SFRCs in the dimensioning process becomes obvious.

Despite the high level of knowledge concerning the experimental analysis of the microstructure and the prediction of mechanical properties using mean-field or full-field methods, there are, however, still some open topics which are treated here. In essence, this work represents a collocation of three publications in peer reviewed journals.

After a brief revision of the common methods to determine microstructural properties of fiber reinforced composites, and the application of mean-field and full-field approaches predicting the effective properties of fiber reinforced composites, possibilities to describe and to visualize the microstructure of SFRCs are discussed in chapter 2. In chapter 3, the homogenization of linear elastic properties using experimentally determined microstructural information is discussed. The predictions of linear elastic properties of mean-field homogenization approaches and full-field voxel-based homogenization methods for short-fiber reinforced materials are opposed in chapter 4. In chapter 5, the description of the microstructure of short-fiber reinforced composites using fiber orientation tensors and their effect on the elastic properties is discussed. Finally, in chapter 6, the main findings of this work are summarized and concluded.

1.2 State of the art

Determination of microstructure properties Information about the microstructure is essential in the prediction of the material behavior by means of micromechanical approaches. Different techniques have been proposed in order to obtain this information (Bernasconi et al., 2012). On the one hand, the microstructure information is determined by means of polished cross-sections. Therefore, the elliptical footprints left by fibers on this polished cross-sections are observed (Fakirov and Fakirova, 1985; Bay and Tucker III, 1992; Clarke et al., 1993). The ellipticity is used to determine the angle between the fiber orientation and the observed cross-section. This method allows to determine the orientation of each single fiber, and is often used due to its simplicity. There are, however, a few inherent disadvantages: firstly, since sectioning of the specimen is necessary, it is a destructive method. Secondly, the fiber orientation information emerging from the elliptical footprints is not unique. Several improvements have been suggested, such as the observation of multiple sections (Clarke et al., 1995; Eberhardt et al., 2001). And thirdly, the orientation distribution of the fibers is not directly correlated with the length distribution.

On the other hand, x-ray observation by means of computed tomography are, meanwhile, a common approach in order to obtain detailed information about the microstructure of composites (Shen et al., 2004; Bernasconi et al., 2008; Ohser and Schladitz, 2009). Micro-computed tomography (μCT) in laboratories or synchrotron tomography have been used not only to observe and evaluate microstructures, but also to characterize damage through *ex situ* and *in situ* observations (Bull et al., 2013, 2014). A detailed review on x-ray tomography can be found in (Maire and Withers, 2013). This methodology produces a three-dimensional voxel-based picture of the specimen in gray-scale. This picture has to be processed further in order to analyze the fiber architecture and to gain the information, which is useful for the prediction of mechanical properties.

Generally, three basic methods may be applied to calculate the effective properties based on the μCT data. First, the voxel picture can be used directly within full-field approaches like finite element methods and fast Fourier transformation methods. With regard to the details of the microstructure, efforts to calculate full-field solutions directly based on CT scans are, however, still handicapped by rough meshes, since the computational time is extremely high for these methods (Düster et al., 2012; Węglewski et al., 2013). Second, methods have been proposed, aiming to extract the mean orientation distribution in terms of orientation tensors (Bernasconi et al., 2008). Relying on these tensors, full-field (Demirci et al., 2011; Duschlbauer et al., 2006) and also mean-field (Müller et al., 2015b) approaches may be applied. The third approach consists of a segmentation of the fibrous microstructure and, eventually, a calculation of the mean orientation distribution based on these segmented data. In this case, the effective properties can be calculated by mean-field and also full-field methods incorporating the segmented data directly or in terms of orientation tensors.

Mean-field homogenization of mechanical properties In context of mean-field theories, which are also called average-field theories, the fields on the macroscale are determined by volume averaging the fields on the microscale. The effective properties are then defined as the relation between the averages of the fields on the microscale. This understanding of mean-field theories is motivated by the fact, that the effective mechanical properties from experiments are relations between volume averages of strain and stress (Hori and Nemat-Nasser, 1999). Extensive overviews concerning the linear and non-linear homogenization are given by Willis (1981), Mura (1987), Ponte Castañeda and Suquet (1997), Torquato (2002), Nemat-Nasser and Hori (1999), Gross and Seelig (2007) and Kanouté et al. (2009).

Generally, the set of mean-field methods can be divided into bounding and estimating methods. The former specify an admissible range of possible effective properties for given microstructural information. Since the first order or simple bounds, known as the Voigt and Reuss bounds (Voigt, 1889; Reuss, 1929), only take into account the volume fractions, they enclose a wide range of admissible effective properties. Provided by Hashin and Shtrikman (1962a,b, 1963), the second-order bounds are based on a variation principle. This approach was extended and applied to anisotropic materials (Böhlke and Lobos, 2014) and also nonlinear material behavior (Talbot and Willis, 1985, 1992), as well. Both extensions of the Hashin-Shtrikman variational principle are based on the introduction of a homogeneous comparison material. A further extension of this concept is referred to as the *Linear Comparison Composite* (LCC) approach (Ponte Castañeda and Suquet, 1997). Here, instead of a homogeneous comparison material a heterogeneous comparison material is optimally chosen in order to apply a linear homogenization procedure by incorporating heterogeneities of piecewise constant inelastic fields.

Contrary to the bounding methods, estimating approaches give approximations of the effective properties. The Mori-Tanaka (Mori and Tanaka, 1973), the self-consistent (Kröner, 1977) (SC), the generalized self-consistent (Christensen and Lo, 1979) and the interaction direct derivative (IDD) (Zheng and Du, 2001) estimates are prominent representatives of such mean-field schemes, which take the interaction of the inclusions into account.

Based on the Eshelby solution (Eshelby, 1957), the general idea of SC is to embed each inclusion in an infinite matrix with the properties of the effective material. Due to this assumption, the SC approach owns an inherently implicit character. The SC method was applied to granular and also to particulate materials with multiple phases and non-linear properties (Budiansky, 1970; Gillman et al., 2013; Miled et al., 2011). The major shortcoming of SC is, that the interaction between the inclusions and the surrounding matrix is not considered directly. This drawback is the main motivation to apply methods like IDD, which are based on the three-phase model. Here, one inclusion interacts with the matrix directly, and with the other inclusions through the effective medium.

Additionally, two-step (TS) methods may be applied to homogenize particulate materials with arbitrarily oriented inclusions (Pierard et al., 2004). Within these methods, the microstructure is decomposed into as many domains as there exist different inclusions. Each domain is homogenized individually, and then, all domains are homogenized to determine the effective properties. This method can be applied to a much larger class of microstructures compared to second-order bounding schemes (Doghri and Friebel, 2005) and also in conjunction with strength prediction (Kaiser and Stommel, 2012).

Full-field homogenization of mechanical properties In the last years, serious advance in the research and application of numerical homogenization techniques based on full-field simulations has been obtained (Suquet, 1987; Guedes and Kikuchi, 1990; Ghosh et al., 1995; Moulinec and Suquet, 1994; Andrä et al., 2013b). This approach allows for the treatment of complicated microstructures, inelastic material laws, e.g., plasticity or viscoplasticity and also multi-physics problems (Hain and Wriggers, 2008; Geers et al., 2010; Roland et al., 2014; Helfen and Diebels, 2014).

For full-field methods, a boundary value problem is defined on a representative volume element (RVE). The microstructure of the composite is represented in the RVE. The local field in the RVE caused by an external load may be determined by, e.g., finite element methods or fast Fourier transformation approaches. Generally, the choice of the size of the RVE, and the definition of proper boundary conditions with regard to the considered microstructure are essential when applying this method (Dirrenberger et al., 2014).

The Fourier method proposed by Moulinec and Suquet (1994) applies the fast Fourier transformation (FFT) to solve integral equations of Lippmann-Schwinger type (Zeller and Dederichs, 1973; Kröner, 1977). These integral equations are equivalent to the local periodic cell problems in numerical homogenization. In this approach meshing and the assembly of the linear system is not required. Thus, the memory needed for the solution of the boundary value problem is significantly smaller compared to other methods. Since very fast FFT implementations (Frigo and Johnson, 1998, 2005; Johnson and Frigo, 2007) are available, this approach has been applied for real word problems (Andrä et al., 2013a). A full-field FFT approach has been compared to various mean-field methods based on artificial variation of the properties of the constituents of a composite consisting of a matrix and spherical inclusions (Ghossein and Lévesque, 2012) and to experimental data (Spahn et al., 2014).

Regarding the high level of knowledge concerning the experimental analysis of the microstructure and the prediction of mechanical properties, the key achievement of this work is the detailed qualitative and quantitative comparison of mean-field, full-field and experimental methods for the case of SFRC by combining established methods from mechanics, applied mathematics and materials science.

1.3 Notation and frequently used symbols

A direct tensor notation is preferred throughout the text. If tensor components are used, then Latin indices are used and Einstein's summation convention is applied. In the following, frequently used symbols, tensor operations, subscripts and superscripts and abbreviations are listed.

Scalars

a	Aspect ratio
a_1, a_2, a_3	First, second and third half axis of an ellipsoid
c	Volume fraction
d	Diameter
d_{RVE}	Side length of an representative volume element (RVE)
E	Young's modulus
l	Length
N	Number of phases/fibers
r	Radius
S	Surface of unit sphere
V	Volume

Vectors and Tensors

$\boldsymbol{a}, \boldsymbol{b}, \boldsymbol{c}, \dots$	Vectors
$\boldsymbol{A}, \boldsymbol{B}, \boldsymbol{C}, \dots$	Tensors of second order
$\mathbb{A}, \mathbb{B}, \mathbb{C}, \dots$	Tensors of fourth order
$\mathbb{A}$	Strain localization tensor
$\mathbb{C}$	Stiffness tensor
$\boldsymbol{D}$	Deviatoric part of second-order orientation tensor
φ	Azimuthal angle of spherical coordinate system
$\boldsymbol{\varepsilon}$	Strain tensor
$\boldsymbol{I}$	Second-order identity
$\mathbb{I}^{\mathsf{s}}$	Symmetric part of fourth-order identity
$\boldsymbol{n}$	Unit vector, fiber axis orientation
$\boldsymbol{N}$	Second-order orientation tensor
ν	Poisson ratio
$\lambda_1, \lambda_2, \lambda_3, \dots$	Eigenvalues
$\boldsymbol{p}$	Stress polarization

$\mathbb{P}$	Hill's polarization tensor
$\boldsymbol{\sigma}$	Stress tensor
$\mathbb{S}$	Compliance tensor
θ	polar angle of spherical coordinate system
$\boldsymbol{u}$	Displacement vector
$\boldsymbol{Z}$	Ellipsoidal description of fiber geometry
$\boldsymbol{Z}^{\mathsf{D}}$	Ellipsoidal description of fiber distribution

Tensor operations

$\langle\cdot\rangle$	Volume/ensemble averaging
$[\cdot]$	Linear mapping of second-order tensors by a fourth-order tensor: $\boldsymbol{A} = \mathbb{C}[\boldsymbol{B}]$
$\cdot$	Scalar product: $\boldsymbol{A} \cdot \boldsymbol{B}$
$\square$	Box product: $(\boldsymbol{A}\square\boldsymbol{B})[\boldsymbol{C}] = \boldsymbol{A}\boldsymbol{C}\boldsymbol{B}$
$\otimes$	Dyadic product: $(\boldsymbol{A} \otimes \boldsymbol{B})[\boldsymbol{C}] = (\boldsymbol{B} \cdot \boldsymbol{C})\boldsymbol{A}$
$[\![\cdot]\!]$	Contraction operator: $(\boldsymbol{a} \otimes \boldsymbol{b}) \cdot (\mathbb{C}[\![\boldsymbol{a} \otimes \boldsymbol{b}]\!]) = (\boldsymbol{a} \otimes \boldsymbol{a}) \cdot (\mathbb{C}[\boldsymbol{b} \otimes \boldsymbol{b}])$
$\boldsymbol{n}^{\otimes\alpha}$	Higher-order dyadic products of the same tensor: $\boldsymbol{n}^{\otimes\alpha} = \boldsymbol{n} \otimes \cdots \otimes \boldsymbol{n}$, $\boldsymbol{n}^{\otimes\alpha}$ is a tensor with the rank α times the rank of $\boldsymbol{n}$
$\mathrm{sym}(\cdot)$	Symmetric part of a tensor
$\mathrm{grad}\,(\cdot)$	Eulerian gradient of a quantity

Subscripts and superscripts

$(\cdot)_0$	Initial or reference quantity
$(\cdot)_{\mathsf{F}}$	Reference to fiber
$(\cdot)_{\mathsf{M}}$	Reference to matrix

$(\cdot)_\alpha$	Reference to inclusion/fiber α
$\bar{(\cdot)}$	Averaged quantity
$(\cdot)^\pm$	Upper and lower bound
$(\cdot)'$	Irreducible part of quantity
$(\cdot)_{\langle\alpha\rangle}$	Tensor of rank α
$\hat{(\cdot)}$	Quantity in Fourier space

Abbreviations

BS	Basic scheme of the FFT solution
DFT	Discrete fast Fourier transformation
DIC	Digital image correlation
FFT	Fast Fourier transformation
FRC	Fiber reinforced composite
HS	Hashin-Shtrikman
IDD	Interaction direct derivative
µCT	Micro-computed tomography
MF	Mean field
PP	Polypropylene
PPGF30	Polypropylene reinforced with 30wt.% of glass fibers
RVE	Representative volume element
SFRC	Short-fiber reinforced composite
SC	Self-consistent
TS	Two-step (mean-field approach)
TP	Thin plate (microstructure type)
UD	Unidirectional (microstructure type)
vol.%	Volume fraction
wt.%	Weight fraction

2 Microstructure of short-fiber reinforced composites

2.1 Classification and general modeling assumptions

Generally, materials with heterogeneous microstructures are modeled as assemblies of a set of N homogeneous phases. The total volume V of the aggregate is composed of all volume parts V_α occupied by the phases $\alpha = 1, 2, \ldots N$:

$$V = \sum_{\alpha=1}^{N} V_\alpha, \quad \alpha \in 1, 2, \ldots N. \tag{2.1}$$

Each of the phases with the volume V_α has certain physical properties, a certain geometrical shape, and a specific orientation. Since there are no limitations in the geometrical and spatial compilation of a heterogeneous microstructure of solid materials, a simplified classification of these microstructures is convenient for this work: a heterogeneous microstructure may be of *particulate* or *granular* type (Agoras and Ponte Castañeda, 2011). On the one hand, materials

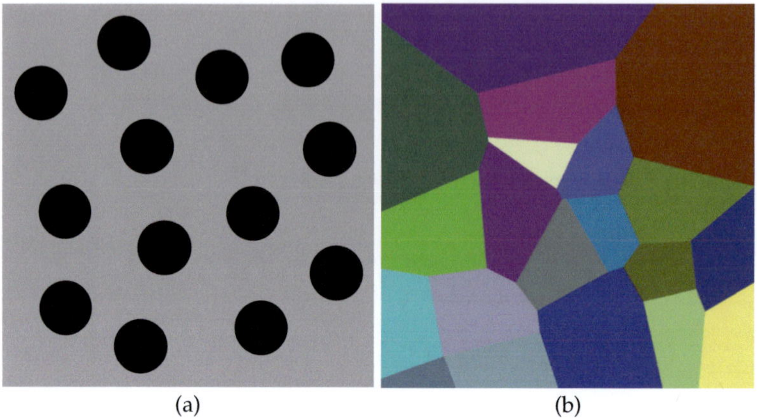

Figure 2.1: Illustrations of (a) a particular and (b) a granular microstructure

with particle or fiber reinforcements are examples for materials with particulate microstructures. This type of materials exhibits a distinct phase. This distinct phase, called *the matrix* in the following, surrounds all the over phases, see Figure 2.1(a). On the over hand, polycrystalline materials represent exemplarily the family of granular materials, see Figure 2.1(b). In both microstructure types, a phase comprises all constituents with similar properties, shape and orientation. Thus, a polycrystalline material can, generally, be referred to as multiphase composite, where each set of anisotropic grains with similar alignment may constitute a particular phase. A typical two-phase composite is, e.g., a matrix-based material, which is reinforced with dispersed and isotropic spherical particles.

Fiber reinforced composites (FRCs) exhibit a particulate microstructure. The main constituent of FRCs is the fiber. Apart from the physical properties, the length is the critical attribute of the fibers. By means of the fiber length, FRCs are divided into two groups: FRCs with long or short fibers are referred to as *continuous* or *discontin-*

uous FRCs, respectively. In addition to the length, the orientation distribution of the fibers plays an important role. Especially in case of continuous FRCs, the fiber orientation can be adjusted properly to meet the requirements of external loads. In case of discontinuous FRCs, the manufacturing process, however, mainly defines the fiber orientation distribution.

In this work, the focus is exclusively on discontinuous, short-fiber reinforced composites (SFRCs). For the fibers, a cylindrical shape with a circular cross section and a constant diameter is assumed. Since the aspect-ratio, i.e., the length to diameter ratio, is moderate, the fibers are assumed to be straight. Both, the matrix and the arbitrary oriented fibers are assumed to be linear elastic and isotropic with piecewise constant properties. The material behavior of the fibers is assumed to be uniform. The theoretical approaches are considered in the concepts of small strains with quasi-static loading and linear elasticity by neglecting viscous effects.

2.2 Mathematical description of microstructures

A heterogeneous microstructure can be regarded as a specific realization of a stochastic process. The set of all realizations or samples of this process is called ensemble. The volume of the ensemble Ω is much larger than the volume of a single sample ω. In each sample, the volume fraction of all phases $c_\alpha = V_\alpha/V$ is constant. The total volume of phase α in sample ω is denoted by $V_\alpha(\omega)$. Thus, the function $\Phi = \Phi(\boldsymbol{x}, \omega)$ describing a specific mechanical property depends on the location in space $\boldsymbol{x}$ and a particular sample ω of the stochastic process. The *ergodicity hypothesis* states that averaging $\Phi(\boldsymbol{x}, \omega)$ over all samples of an ensemble at a constant position $\boldsymbol{x}_0$ delivers the same result as averaging this function over one realization ω_0 with an infinitely large volume:

$$\langle \Phi(\boldsymbol{x}_0, \omega) \rangle_\Omega = \langle \Phi(\boldsymbol{x}, \omega_0) \rangle_{\omega_0} \quad \omega_0 \rightarrow \infty, \tag{2.2}$$

where $\langle \cdot \rangle_\Omega$ denotes the ensemble averaging and $\langle \cdot \rangle_{\omega_0}$ the averaging over the volume ω_0. Since usually a few microstructure realizations are available only it is possible to calculate, e.g., the mechanical properties using volume averaging for ergodic materials. For this purpose, the indicator function, also referred to as characteristic function, is defined as

$$\mathcal{I}_\alpha(\boldsymbol{x}, \omega) = \begin{cases} 1 & \forall \boldsymbol{x} \in V_\alpha(\omega) \\ 0 & \text{otherwise} \end{cases} . \tag{2.3}$$

The indicator function $\mathcal{I}_\alpha(\boldsymbol{x}, \omega)$ yields 1 if the position vector $\boldsymbol{x}$ depicts a point in phase α within the sample ω, and 0 in all other cases. For the sake of completeness, it should be mentioned that this function is complete, idempotent and bi-orthogonal:

$$\sum_\alpha^N \mathcal{I}_\alpha(\boldsymbol{x}, \omega) = 1 \qquad \text{(complete)}, \tag{2.4}$$

$$\mathcal{I}_\alpha(\boldsymbol{x}, \omega)\mathcal{I}_\alpha(\boldsymbol{x}, \omega) = \mathcal{I}_\alpha(\boldsymbol{x}, \omega) \qquad \text{(idempotent)}, \tag{2.5}$$

$$\mathcal{I}_\alpha(\boldsymbol{x}, \omega)\mathcal{I}_{\beta \neq \alpha}(\boldsymbol{x}, \omega) = 0 \qquad \text{(bi-orthogonal)}. \tag{2.6}$$

The ensemble average of the indicator function is denoted by the *one-point probability function* of phase α:

$$S_1^\alpha(\boldsymbol{x}) = \langle \mathcal{I}_\alpha(\boldsymbol{x}, \omega) \rangle_\Omega. \tag{2.7}$$

This function gives the probability to meet phase α at the location $\boldsymbol{x}$ in the ensemble Ω. In the case of homogeneous material, this function is independent of the position vector $\boldsymbol{x}$ and gives the volume fraction c_α of the corresponding phase. In the case of heterogeneous materials, $S_1^\alpha(\boldsymbol{x})$ can be regarded as the position dependent volume fraction of phase α. Higher-order correlation functions (n-point correlations) are defined in a similar way. Since these functions do not play a role in this work, the reader may refer to Ohser and Mücklich (2000); Torquato (2002) and Dvorak (2012) for a detailed evaluation of these functions.

Along with the ergodicity hypothesis, the *statistical homogeneity* and *statistical isotropy* are, however, important properties of microstructures. For statistical homogeneous microstructures, the n-point correlation functions are translation invariant. If the n-point correlation functions are, additionally, rotational invariant, then the microstructure is referred to as statistical isotropic. In all, for statistical isotropic microstructures, the one-point correlation function is constant and equal to the volume fraction of the corresponding phase, and the higher-order correlations only depend on a scalar distance.

2.3 Convenient description of microstructure of SFRCs

In this work, the microstructure of SFRCs is considered in two different ways: firstly, a discrete description and, secondly, an averaged description are used. Within the discrete description, each fiber is fully characterized as cylinder with the position vector $\boldsymbol{p}_\alpha$,

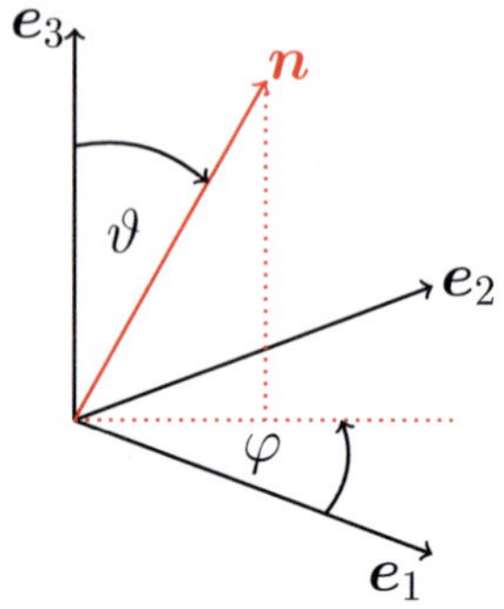

Figure 2.2: Parametrization of the fiber axis $\boldsymbol{n}$ in spherical coordinates

the axis orientation vector $\boldsymbol{n}_\alpha$, the length l_α, and the radius r_α. Additionally, the aspect ratio $a_\alpha = l_\alpha/(2r_\alpha)$ of the fiber can be computed as the ratio of the length to the diameter. The axis of the fiber is parameterized in spherical coordinates with the angles ϑ and φ, see Figure 2.2. The total fiber volume fraction is indicated with $c_\mathsf{F} = 1 - c_\mathsf{M} = \sum_{\alpha=1}^{N} c_\alpha$, where c_M denotes the volume fraction of the matrix, c_α the volume fraction of fiber α, and N the number of different fibers. In chapters 3 and 4, this microstructure description is applied.

In chapter 5, the average description is used. This approach relies on an average volume fractions of fibers c_F and matrix c_M, mean aspect ratio $\bar{a}$ respectively mean length $\bar{l}$, and a mean description of the orientation distribution of the fibers using tensorial quantities called orientation tensors or fiber orientation distribution functions (FODFs). Details concerning orientation tensors and FODFs are given in chapter 5.

2.4 Visualization of orientation distributions and scalar valued direction dependent functions

In materials science, the orientation distribution of crystallographic lattice planes is often represented by pole figures. A pole figure shows the orientation of an object in space. It is a stereographic projection of poles. Whereby, a pole is the intersection point of a sphere and a line going through its origin. Pole figures are convenient to represent discrete orientations, such as discrete fiber orientations $\boldsymbol{n}_\alpha$, and also scalar valued directional dependent functions $f(\boldsymbol{n})$.

In Figure 2.3(a), the construction of a pole figure is shown, as it is used in this work. Here, a unit sphere is divided in an upper and a lower hemisphere by a reference plane called the equatorial plane. The south pole P_{south} of the unit sphere is used as reference pole for the projection of the intersections P of the orientation axes $\boldsymbol{n}$, with the upper hemisphere on the equatorial plane P'.

In Figure 2.3(b), the equatorial plane and the projection point P' are shown. Generally, stereographic projections of this kind are conformal but not isometric, meaning that the angle between two projected points is preserved, however, the distance not. Thus, in case of a random distribution of orientations, the intersection points on the unit sphere would be uniformly distributed. The pole figure would show an accumulation of points in its center, see Figure 2.5(a).

A direct projection of single orientations as points on the equatorial plane is meaningful as long as not too many orientations have to be

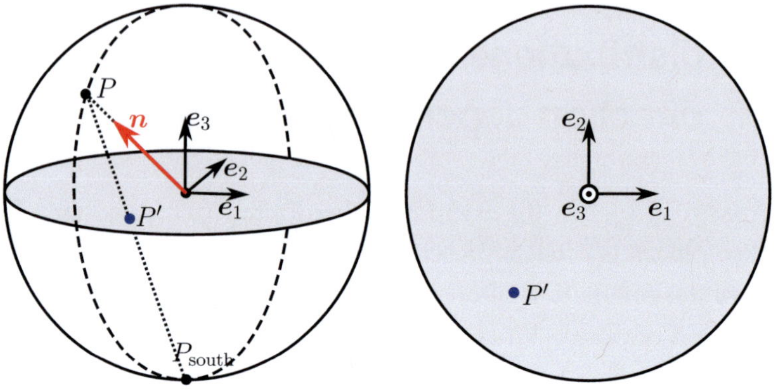

Figure 2.3: (a) Principle construction of a pole figure; (b) Illustration of a pole figure for one orientation

shown. Otherwise, a continuous representation of the pole figure is appropriate. Therefore, a unimodal kernel function $k\left(\boldsymbol{n}, \boldsymbol{n}_0, \kappa\right)$ is chosen, which is radially symmetrical with the rotation axis $\boldsymbol{n}_0$ and the parameter κ controlling the localization of the kernel function around $\boldsymbol{n}_0$. This is a scalar valued and directional dependent function. Considering a set of N discrete fiber orientations, the continuous distribution function $\tilde{f}(\boldsymbol{n})$, which is used to create the pole figure, is defined as

$$\tilde{f}\left(\boldsymbol{n}\right) = \sum_{\alpha=1}^{N} \psi_\alpha k(\boldsymbol{n}, \boldsymbol{n}_\alpha, \kappa). \tag{2.8}$$

Here, ψ_α are weighting factors, which may be used to take the volume fractions of the fibers into account. If all orientations have an equal probability, $\psi_\alpha = \psi = 1/N$ holds. An appropriate kernel function in the analysis of fiber orientations is, e.g., the von Mises-Fisher distribution function. Since fiber orientation distributions are axial

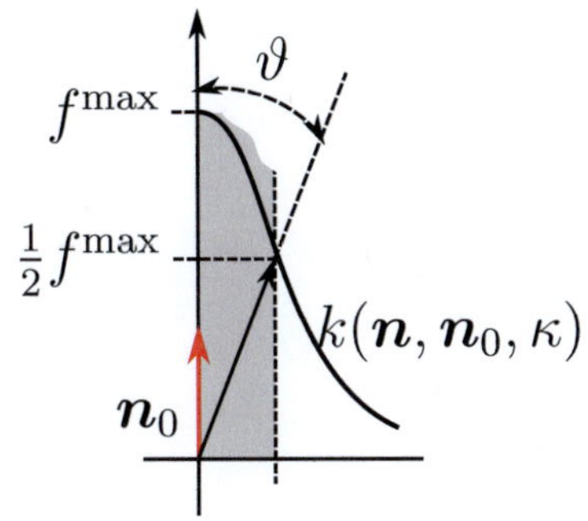

Figure 2.4: Determination of the concentration parameter κ

distributions, the von Mises-Fisher distribution (Mardia and Jupp, 2009) is symmetrized :

$$k^{\mathrm{vMF}}\left(\boldsymbol{n}, \boldsymbol{n}_0, \kappa\right) = \frac{\kappa}{2\sinh\left(\kappa\right)}\left(\exp\left(\kappa \boldsymbol{n}_0 \cdot \boldsymbol{n}\right) + \exp\left(-\kappa \boldsymbol{n}_0 \cdot \boldsymbol{n}\right)\right), \quad (2.9)$$

where $\boldsymbol{n}_0$ is the reference direction and κ the concentration parameter. This function is radially symmetric with respect to $\boldsymbol{n}_0$. It describes a uniform distribution if $\kappa = 0$. In the construction of $\tilde{f}(\boldsymbol{n})$, κ is chosen in such a way that the half width at half maximum is located at an angle ϑ from the reference direction, see Figure 2.4. Once κ is calculated, $\tilde{f}(\boldsymbol{n})$ emerges from a superposition of the kernel function for each orientation according to the prescription in equation (2.8), where the reference direction $\boldsymbol{n}_0$ in the kernel function coincides with the fiber orientations $\boldsymbol{n}_\alpha$. Then, the resulting directional dependent function $\tilde{f}(\boldsymbol{n})$ can be visualized by means of a pole figure.

In Figure 2.5, a discrete and a continuous pole figure for a uniform and a transversal symmetric orientation distribution of equally weighted fiber orientations, respectively, are shown. In case of the continuous pole figure, the half width at half maximum is located at $\vartheta = 15°$.

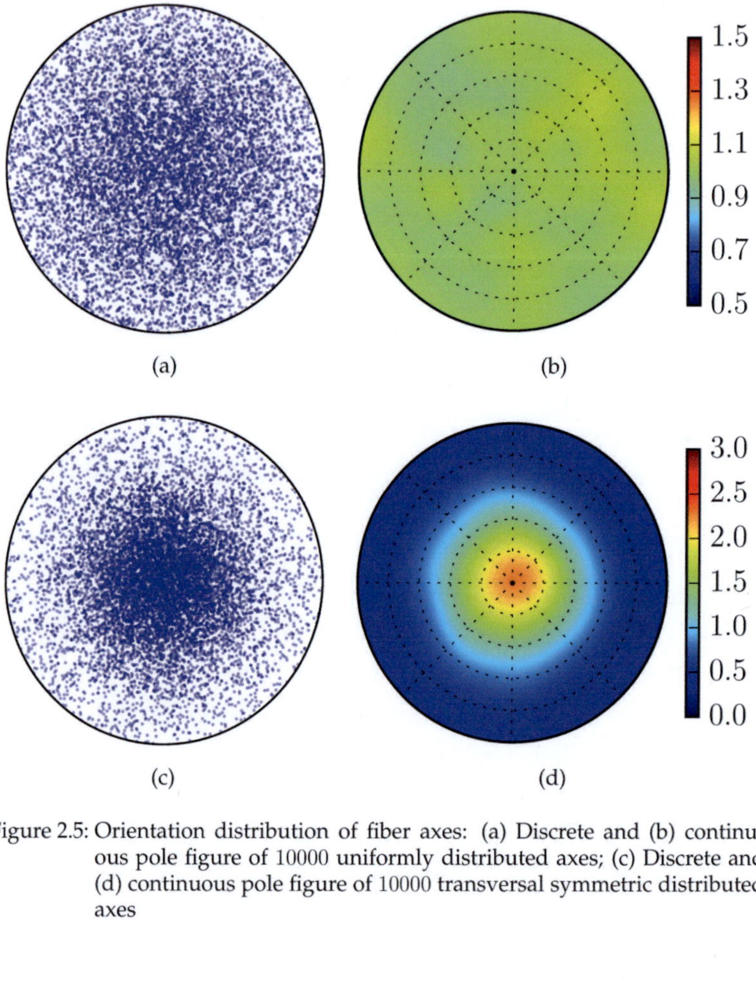

Figure 2.5: Orientation distribution of fiber axes: (a) Discrete and (b) continuous pole figure of 10000 uniformly distributed axes; (c) Discrete and (d) continuous pole figure of 10000 transversal symmetric distributed axes

3 Homogenization of elastic properties based on measured microstructure data[1]

3.1 Introduction

In this chapter, a composite material consisting of polypropylene reinforced with 30wt.% of short glass fibers (PPGF30) is examined. The microstructure of this composite material is analyzed and segmented using μCT and a segmentation algorithm. The segmented microstructure data consist of non-aligned fibers with varying aspect-ratios. It is used within three homogenization approaches directly, namely the SC method, the IDD and a bounding TS approach. In all applied methods, each fiber is considered in the homogenization procedure without a need of orientation averaging of transversal isotropic stiffnesses or compliances for unidirectional structures like it is often done in conjunction with the Mori-Tanaka approach (Ghossein and Lévesque, 2014).

[1] This chapter is based on the paper "Homogenization of elastic properties of short-fiber reinforced composites based on measured microstructure data" (Müller et al., 2015a)

3.2 Experimental methods

3.2.1 Fabrication of specimens

The materials for the test specimens have been compounded at Fraunhofer LBF[2]. Thus, full control over all constituents has been achieved. The polypropylene (PP) Moplen HP500N (Lyondell-Basell) has been used as matrix material. Along with a fraction of 30wt.% of short glass fibers, three additives have been added to the polymeric bulk material in order to assure stability, durability and an adequate performance of the reinforced material. A fraction of 0.5wt.% of Irganox B225 (BASF) served as long-term thermal stabilizer, and additionally, 0.07wt.% of the calcium stearate acid scavenger Ceasit AV (Baerlocher) and 3.0wt.% of the coupling agent Exxelor PO 1015 (Exxon Mobil) have been added. These additives have also been added to the pure PP material. Thus, the effect of the reinforcing fibers can be determined.

Both compounded material, pure PP and the composite PPGF30, have been used to manufacture plates of the dimensions $80\,\mathrm{mm} \times 80\,\mathrm{mm} \times 2.5\,\mathrm{mm}$ by means of injection molding. A triangular gating system has been chosen as infeed. Thus, a homogeneous filling of the plate cavity with a parallel melt front has been achieved (Becker, 2009). From these plates, the specimens for the mechanical testing have been machined. In case of the pure PP material, the specimens have been extracted at $0°$ and $90°$ in reference to the filling direction during injection molding. In case of PPGF30, specimens with three orientations ($0°$, $45°$ and $90°$) have been prepared. Hence, the anisotropic material behavior can be determined.

[2] The preparation of the specimens and the tensile tests have been performed by F. Dillenberger (Fraunhofer LBF).

Details concerning the specimen geometry of the tension rod as well as the different extraction angles are shown in Figure 3.1(a).

3.2.2 Quasi-static tensile tests

The quasi-static tensile tests were conducted on a Z020 testing machine by Zwick-Roell with a velocity of 1 mm/min. During the testing procedure, temperature and humidity have been kept constant (23°C and 50%). All samples have been tested within a two week time frame. Thus, influences of long-term post-crystallization processes in the PP matrix have been prevented.

The deformation of the specimen has been observed using 2D digital image correlation (DIC) (Tarigopula et al., 2008; Palanivelu et al., 2009). Different stages of the sample deformation have been captured with a monochromatic CCD-camera equipped with a resolution of 1280 px × 1024 px, a color depth of 8 bit and a pixel-size of 8 µm/px. Subsequently, local strain measurement has been performed using the software package Vic2D (Limes GmbH). This software calculates the strain field of the gray-scale pattern for an area of interest by analyzing the deformation of predefined gray-scale facets between consecutive deformation steps and a reference image. As the correlation results refer to pixel differences, they have to be calibrated to the actual geometry of the specimen. In order to get stress-strain curves, the local strain values in a rectangular area of 12 mm × 2 mm over the region of maximum deformation have been averaged and combined with the data from the machine integrated force sensor. The surface strain field and the averaging area are shown in Figure 3.1(b).

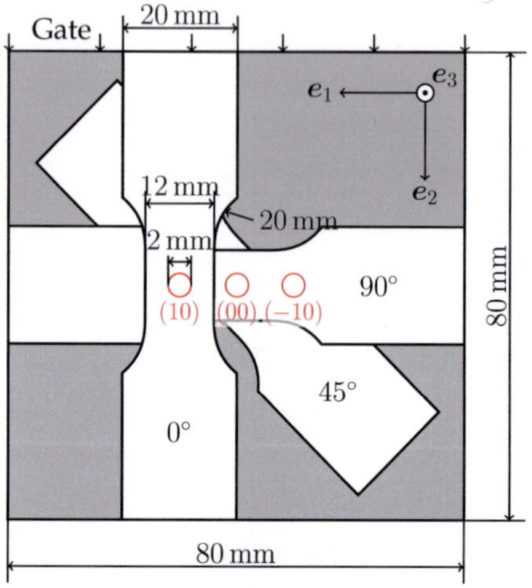

(a)

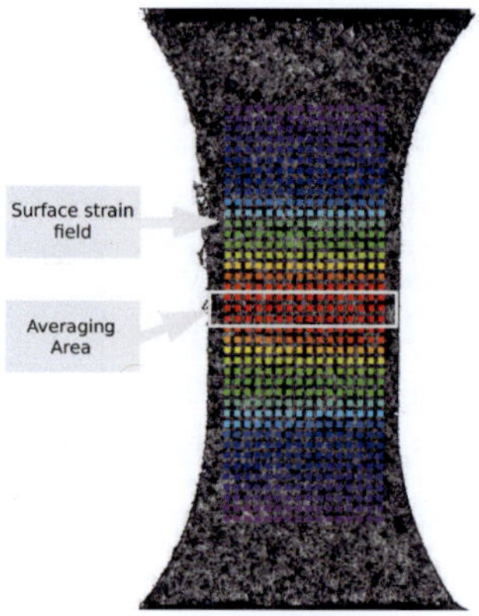

(b)

Figure 3.1: (a) Locations of specimen extraction from plate (circles distinguish position of samples for microstructural analysis), (b) Example of 2D surface strain field calculated by the software Vic2D (Limes GmbH) using digital image correlation (DIC), and averaging area on a tension specimen

The final result has been calculated by averaging values of five single tests for each configuration. In Figure 3.2(a) these results are shown for the pure PP matrix material for specimens oriented in 0° and 90°-direction. Since the difference between the two curves is in the range of scatter of the single experiments, the matrix material has been considered to be isotropic and the results of both orientations have been averaged.

The stress-strain response of the SFRC shows a significant deviation depending on the orientation of sample extraction from the sample plate, see Figure 3.2(b). This mechanical anisotropy is a result of the material orientation induced by the fiber distribution inherent in the injection-molded plates.

The linear elastic properties for the subsequent homogenization procedure have been calculated according to DIN EN ISO 527 1-4 (2012). Since no distinct definition of a linear elastic region is possible, the data has been evaluated in the small strain intervals 0.05%-0.25%, 0.05%-0.5% and 0.05%-1.0%, as shown in Tab. 3.1. Any influence of viscous material behavior has been neglected. Young's modulus has been calculated as the secant gradient in the examined strain range, as shown in 3.2(a). The values of Young's modulus and Poisson's ratio for both materials are given in Tab. 3.1. The Poisson's ratios have been calculated as the secant gradient in the interval of interest of the lateral strain versus longitudinal strain curve. The elastic properties of the glass fibers have been taken from literature and are also shown in Tab. 3.1.

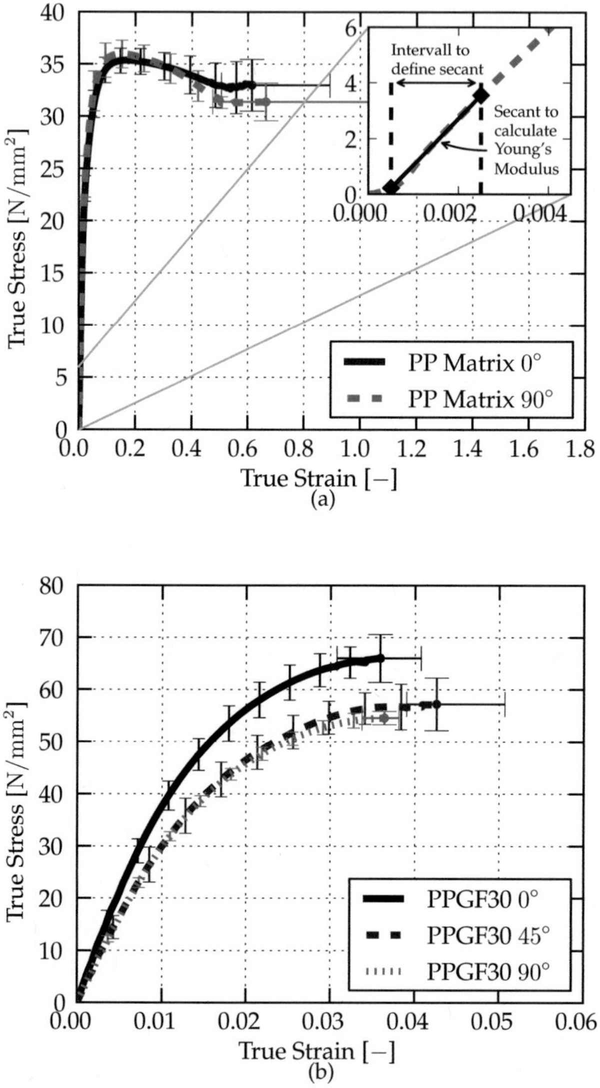

Figure 3.2: (a) Stress-strain curve of PP material showing details on derivation of Young's modulus, (b) Stress-strain curve of PPGF30

3.3 Analysis of microstructure

3.3.1 Micro-computed tomography

The cylindrical specimens for micro-computed tomography (μCT) measurements[3] have been milled from the injection-molded plate at the positions defined in Figure 3.1(a). In order to resolve the reinforcing fibers accurately with the maximal resolution, the diameter of the specimen should have a maximal diameter of 4 mm.

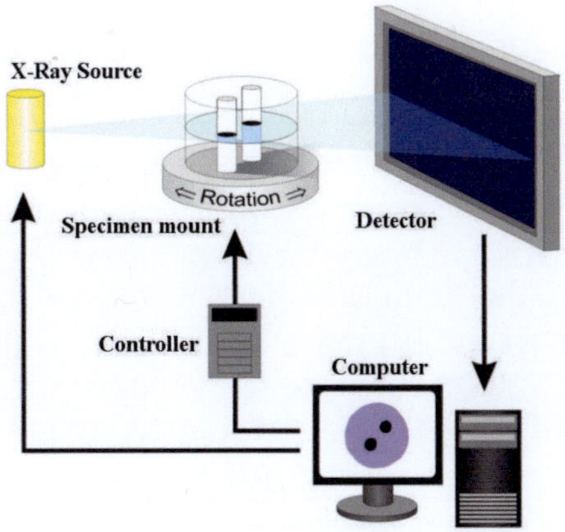

Figure 3.3: μCT measurement setup

The samples have been analyzed with a resolution of 1.8μm. The procedure of a μCT-measurement is shown in Figure 3.3: the rotating specimen is exposed to a low intensity x-ray. Since the absorption of the x-ray depends on the density of the material of the sample, the amount of the x-ray reaching the detector varies with

[3] The μCT measurements have been performed by F. Dillenberger (Fraunhofer LBF).

Table 3.1: Experimental values of Young's modulus E and Poisson's ratio ν

	Strain interval (0.0005) - (0.0025)	
Material	E [MPa]	ν [−]
PP 0°	1665.5 ± 6.6 %	0.364 ± 6.4 %
PP 90°	1748.4 ± 15.3 %	0.346 ± 3.0 %
mean PP	1706	0.355
PPGF30 0°	4482.4 ± 2.9 %	0.271 ± 11.2 %
PPGF30 45°	3540.9 ± 5.4 %	0.304 ± 4.7 %
PPGF30 90°	3452.4 ± 2.2 %	0.217 ± 2.1 %
	(0.0005) - (0.005)	
PP 0°	1608.1 ± 3.3 %	0.360 ± 4.4 %
PP 90°	1663.7 ± 6.0 %	0.357 ± 3.4 %
mean PP	1636	0.359
PPGF30 0°	4137.4 ± 4.8 %	0.273 ± 10.1 %
PPGF30 45°	3250.6 ± 5.1 %	0.300 ± 7.4 %
PPGF30 90°	3227.7 ± 1.8 %	0.222 ± 3.5 %
	(0.0005) - (0.01)	
PP 0°	1450.4 ± 2.1 %	0.371 ± 2.0 %
PP 90°	1501.3 ± 3.1 %	0.368 ± 2.2 %
mean PP	1476	0.37
PPGF30 0°	3671.4 ± 2.9 %	0.294 ± 6.3 %
PPGF30 45°	2947.7 ± 4.3 %	0.310 ± 8.0 %
PPGF30 90°	2938.8 ± 1.2 %	0.238 ± 5.0 %
Glass fibers	73000	0.22

the sample orientation. For each full rotation, the sample is analyzed on the vertical axis. This layer-wise voxel information of the density distribution is transformed to slice images as shown in Figure 3.4 and to a three-dimensional geometrical representation of the sample.

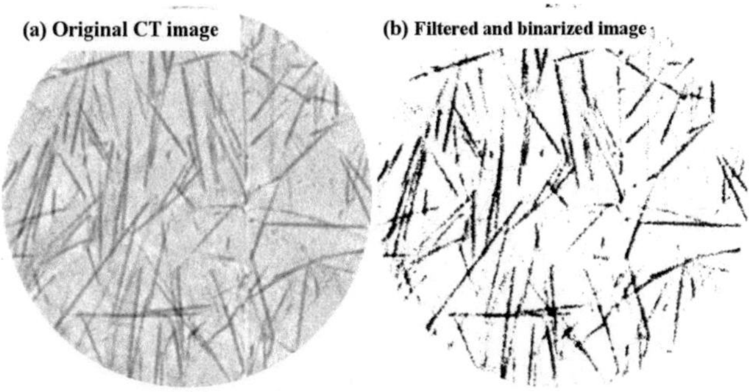

Figure 3.4: μCT slice image of PPGF30

3.3.2 Evaluation of microstructure

The applied evaluation method[4] allows to determine the fiber length, the diameter, the orientation, and the local position distributions by means of an iterative analysis of μCT images (Glöckner et al., 2013). This method is based on model assumptions considering geometrical and symmetrical properties of the fibers: a constant diameter and a small curvature without turning points are assumed. These assumptions allow for derivation of heuristic approaches for the identification of single fibers, and, for assigning of voxels of the μCT image to individual fibers in spite of the inherent image noise. The four steps of the μCT-evaluation process are

[4] The evaluation of the μCT-images has been perfomed by R. Glöckner (Fraunhofer LBF).

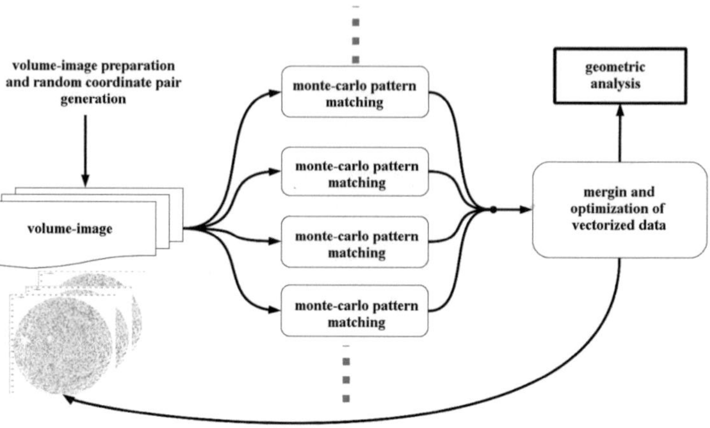

Figure 3.5: Fiber identification process

shown in Figure 3.5. In the first step, the µCT images are filtered and binarizated (see Figure 3.4). Herein, the fiber volume fraction has to be defined in order to achieve a reasonable binarization. For this work, the fiber volume fraction known from the plate fabrication process, has been confirmed by a thermogravimetric analysis. In the second step, the iterative Monte-Carlo pattern identification process is applied. Here, several identification processes are started. During each process, it is tried to identify the center of a separate fiber section. This is done by evaluation integrals over a spherical region at random positions. Then, the fiber orientation n is approximately determined by using the main eigenvalues of second-order moments. Calculating integrals over a cylindrical region for different directions, the exact orientation and length of the fiber is obtained.

The identification process is followed by the control process. Herein, the information of each identification process is merged, voxels of reasonably well detected fibers are deleted from the input image, duplicate fibers are found, and the next iteration of the identification process is started. If a predefined detection accuracy is reached, the control process terminates the identification procedure. In the last step of the µCT-evaluation process, the geometry of the detected fibers is analyzed. The resulting data consisting of the position, orientation, length and diameter of each recognized fiber is called segmented µCT data in the following and can be used to reconstruct the microstructure, see Figure 3.6.

The algorithm has been evaluated in three ways: Firstly, artificially generated fiber data with three different lengths has been mapped on a regular mesh, which corresponds to a 3D voxel picture. The fiber positions have been chosen randomly, however, the length and orientation were correlated. An analysis of this volume-image reveals an overestimation of the fiber length of up to 8% for the short fibers. Secondly, the polymer material of the composite has been removed via combustion and the residual fibers were measured via microscope. Qualitatively similar length distributions have been found. Thirdly, a visual comparison of the volume-image and the detected fibers has been undertaken. This confirmed visually the mentioned small overestimation of the fiber length. Additionally, the reproducibility of the algorithm has been checked by repeated measurement of three different specimen. A standard deviation of the length of approximately 5% has been found. Further details on the image analysis method and its properties can be found in Glöckner et al. (2013).

(a) (b) (c)

Figure 3.6: Segmented μCT data for specimen at the positions (a) (−10), (b) (00) and (c) (10)

3.3.3 Properties of microstructure

In the reconstructions of the microstructure of the evaluated specimen shown in Figure 3.6, the characteristically graded mesostructure of injection-molded plates made of SFRC can be recognized. This phenomenon is well-known in literature (Mlekusch et al., 1999).

Since the segmented μCT data allows a detailed analysis of the properties of the microstructure, the μCT datasets were partitioned into 20 equal-sized layers in through-thickness direction of the plate. The mean aspect-ratio of the fibers $\bar{a}_\gamma$ and the mean orientation distribution in terms of the second-order moment tensor $\boldsymbol{N}_\gamma$ of the fiber orientations have been determined for each layer.

The second-order moment tensor, also called fabric tensor (Kanatani, 1984) or orientation tensor (Advani and Tucker III, 1987), is generally defined by

$$\boldsymbol{N} = \int_S f(\boldsymbol{n})\, \boldsymbol{n} \otimes \boldsymbol{n} \, \mathrm{d}S. \tag{3.1}$$

Herein, $f(\boldsymbol{n})$ is the fiber orientation distribution function, $\boldsymbol{n}$ the fiber axis orientation and $\mathrm{d}S$ a surface element of the unit sphere $S := \{\boldsymbol{n} \in \mathbb{R}^3 : \|\boldsymbol{n}\| = 1\}$. Utilizing Dirac's delta distribution $\delta(\boldsymbol{n} - \boldsymbol{n}_\beta)$ on unit vectors, the empirical orientation distribution function (Kanatani, 1984) for N orientations is given by

$$f(\boldsymbol{n}) = \frac{1}{N} \sum_{\beta=1}^{N} \delta(\boldsymbol{n} - \boldsymbol{n}_\beta). \tag{3.2}$$

Inserting the last equation in the definition of the second-order tensor given in equation (3.1), results in a discrete form of the second-order fabric tensor:

$$\boldsymbol{N} = \frac{1}{N} \sum_{\beta=1}^{N} \boldsymbol{n}_\beta \otimes \boldsymbol{n}_\beta. \tag{3.3}$$

Since the available microstructure data does not contain only the orientations of the fiber axes, but also their length and diameter, it is reasonable to consider these quantities during the calculation of the fabric tensor. Thus, in this work, fabric tensors are considered in a weighted manner. For each of the 20 layers, the weighted fabric tensor $\boldsymbol{N}_\gamma$ is calculated as

$$\boldsymbol{N}_\gamma = \sum_{\beta=1}^{N_\gamma} \omega_\beta \boldsymbol{n}_\beta \otimes \boldsymbol{n}_\beta. \tag{3.4}$$

In the last equation, N_γ is the number of fibers in layer γ, and ω_β is the volume fraction of the fiber with the axis direction $\boldsymbol{n}_\beta$.

In the figures 3.7, 3.8 and 3.9, the mean aspect-ratio of the fibers $\bar{a}_\gamma$, the diagonal components (N_{11}, N_{22}, N_{33}), and the eigenvalues (λ_1, λ_2, λ_3) of the second-order fabric tensor $\boldsymbol{N}_\gamma$ are shown for all three segmented µCT datasets. The indices $(\cdot)_{11}$, $(\cdot)_{22}$ and $(\cdot)_{33}$ correspond to the $\boldsymbol{e}_1$, $\boldsymbol{e}_2$ and $\boldsymbol{e}_3$-direction, respectively, introduced in Figure 3.1(a), whereby, the $\boldsymbol{e}_2$-direction is the filling direction.

From the figures 3.7(a), 3.8(a) and 3.9(a) it is obvious, that the mean aspect-ratio of the fibers in each layer is not constant throughout the thickness of the plate. Especially in the case of the µCT-specimen (-10) and (00), see Figure 3.1(a), the mean aspect-ratio of the fibers in the layers near the bounding sections are slightly larger than those in the core section. The overall mean aspect-ratio and the overall mean length for each specimen is approximately 26.3 and 330 µm, respectively. The overall mean aspect-ratio is marked in the figures 3.7(a), 3.8(a) and 3.9(a).

The fiber orientation distribution is also not constant throughout the thickness. In the figures 3.7(b), 3.8(b) and 3.9(b), the diagonal components N_{11}, N_{22} and N_{33}, and in the figures 3.7(c), 3.8(c) and 3.9(c), the eigenvalues of the fabric tensor $\boldsymbol{N}_\gamma$, are shown. In particular, the former three diagrams highlight the sectional structure of injection-molded plates made of SFRCs: In the bounding sections, layers 1-2 and 19-20, respectively, the majority of the fibers is oriented in the filling direction. In the layers 11-12 at the core section of the specimen, the fibers are oriented in the transversal direction.

The fiber orientation distribution is often characterized only through the diagonal components N_{11}, N_{22} and N_{33} of the second-order fabric tensor $\boldsymbol{N}_\gamma$ (e.g., Laspalas et al. (2008), Vincent et al. (2005)). The comparison of the diagonal components with the eigenvalues of $\boldsymbol{N}_\gamma$ of the 8th and 14th layer for the (00) data in Figure 3.8(b) and (c) highlights, that the diagonal components do not specify completely the orientation distribution. According to the diagonal components of $\boldsymbol{N}_\gamma$ in the aforementioned layers, the orientation distribution seems to be planar isotropic. The eigenvalues for the same cases indicate, however, a planar isotropic distribution for layer 14, not for layer 8.

These results are confirmed through the stereographic projections of all fiber axes orientations in each of the layers 8, 11 and 14 in Figure 3.10. Here, the out-of-plane direction coincides with the filling direction ($\boldsymbol{e}_2$-direction) and the horizontal direction with the transversal direction ($\boldsymbol{e}_1$-direction). Thus, the vertical direction in these figures corresponds to the through thickness direction, which is perpendicular to the injection-molded plate.

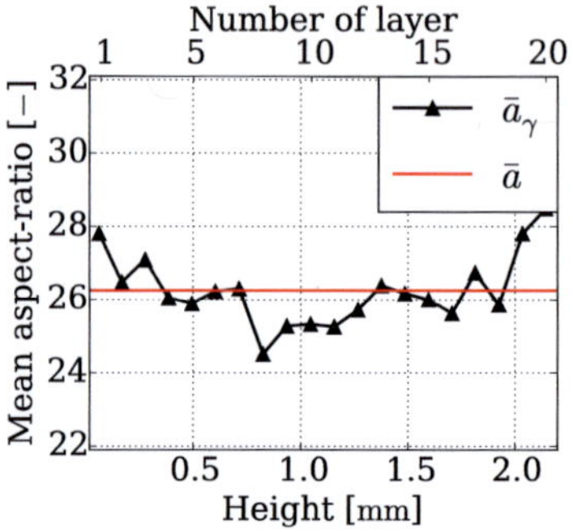

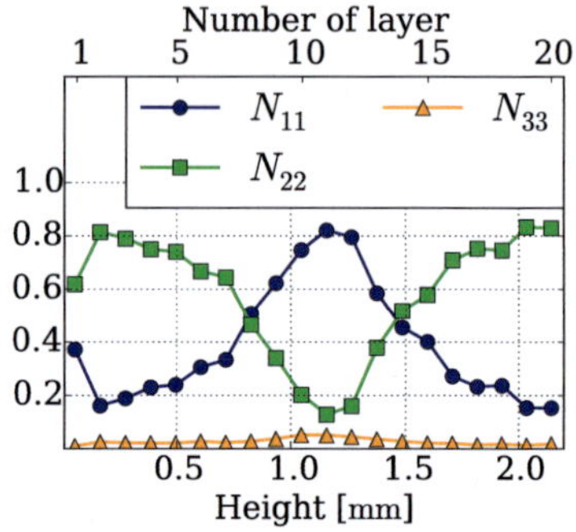

(a) Mean aspect-ratio of the fibers in each layer $\bar{a}_\gamma$ and the whole dataset $\bar{a}$

(b) Diagonal components of $\boldsymbol{N}_\gamma$ for each layer

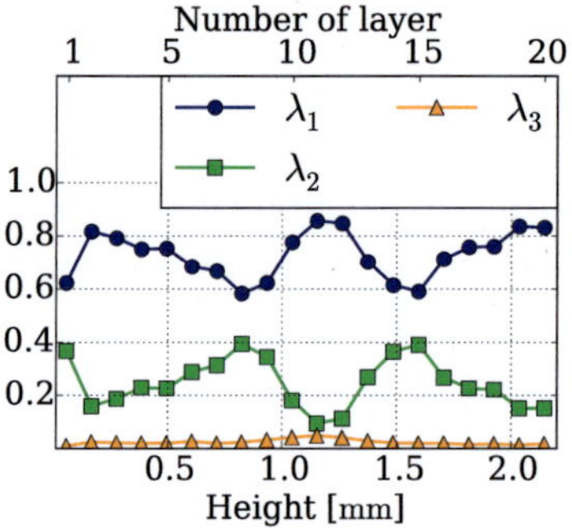

(c) Eigenvalues of $\boldsymbol{N}_\gamma$ for each layer

Figure 3.7: Microstructural properties of the μCT specimen at (-10)

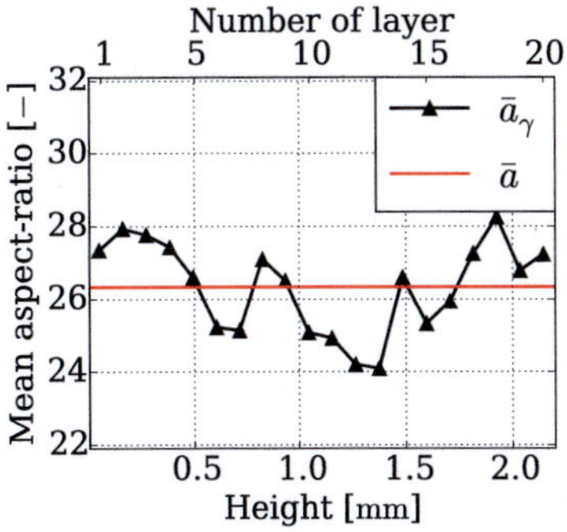

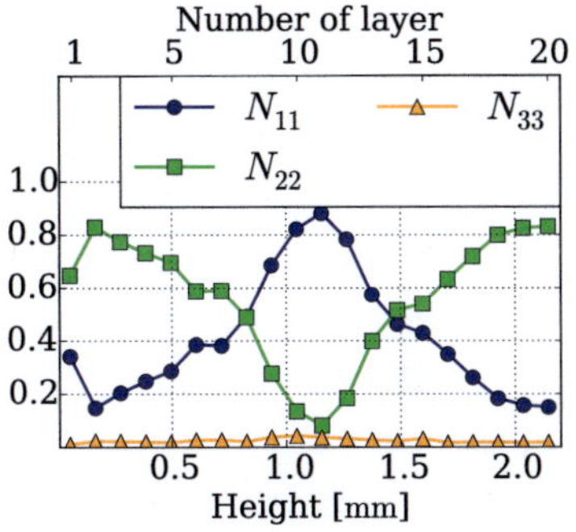

(a) Mean aspect-ratio of the fibers in each layer $\bar{a}_\gamma$ and the whole dataset $\bar{a}$

(b) Diagonal components of $\boldsymbol{N}_\gamma$ for each layer

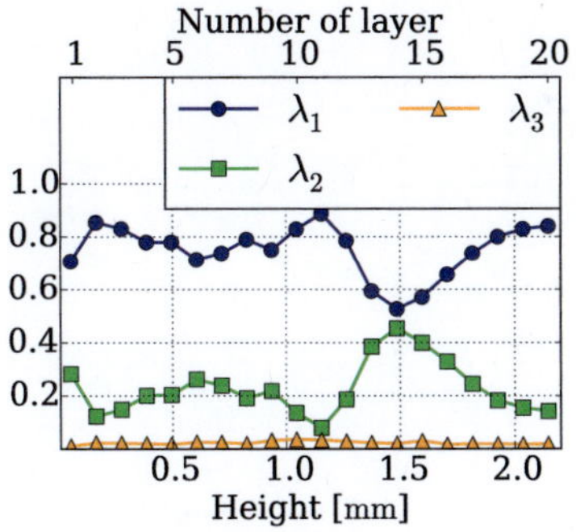

(c) Eigenvalues of $\boldsymbol{N}_\gamma$ for each layer

Figure 3.8: Microstructural properties of the μCT specimen at (00)

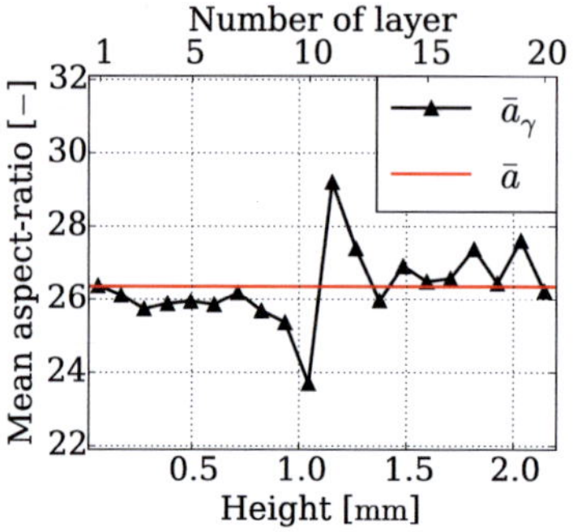

(a) Mean aspect-ratio of the fibers in each layer $\bar{a}_\gamma$ and the whole dataset $\bar{a}$

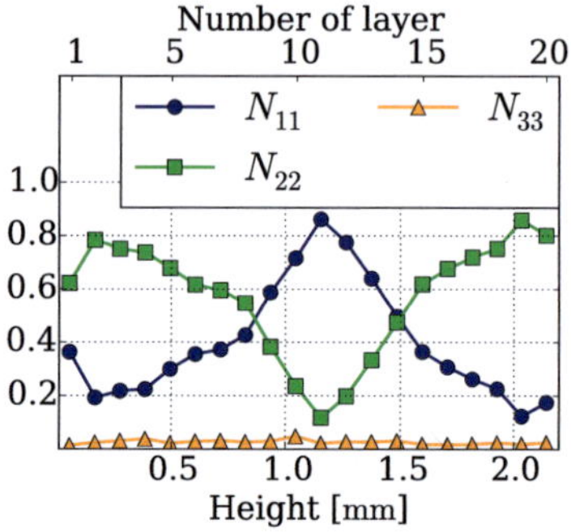

(b) Diagonal components of $\boldsymbol{N}_\gamma$ for each layer

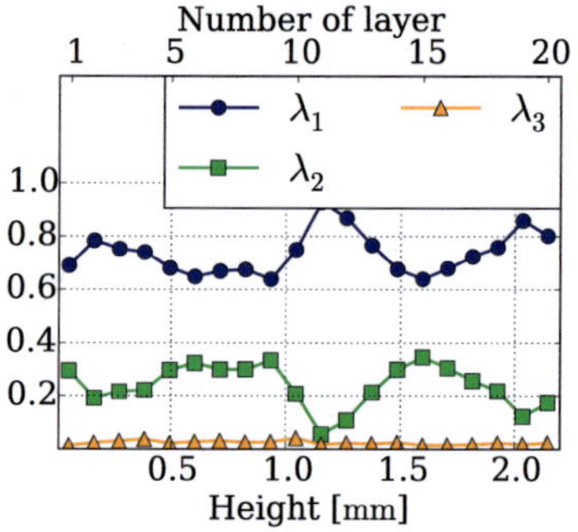

(c) Eigenvalues of $\boldsymbol{N}_\gamma$ for each layer

Figure 3.9: Microstructural properties of the μCT specimen at (10)

In Figure 3.10(a), a preferred orientation of the fibers between the out-of-plane direction and the transversal direction in layer 8 can be seen. In the pole figure for layer 11, Figure 3.10(b), the preferred orientation of the fibers corresponds to the transverse direction. In the case of layer 14 in Figure 3.10(c), an approximately planar isotropic fiber orientation distribution can be noticed. The through-thickness orientation distribution in the other segmented µCT datasets is qualitatively and quantitatively similar to the section discussed before.

The stereographic projections of the entire segmented µCT data are shown in Figure 3.11. Since each µCT dataset contains more than 6000 fibers, continuous representations of the stereographic projections are shown. In all examples the fiber orientation distribution is only approximately planar isotropic. There exists rather a preferred orientation different from the filling direction. This can be affiliated with the manufacturing process and the flow conditions during manufacturing. The mean lengths $\bar{l}$, the diagonal components and the eigenvalues of $\boldsymbol{N}$ for the entire datasets are summarized in Tab. 3.2.

3.4 Homogenization of linear elastic properties

3.4.1 Modeling preliminaries

In the present work, the polymeric matrix and the arbitrary oriented glass fibers are assumed to be linear elastic and isotropic with piece-

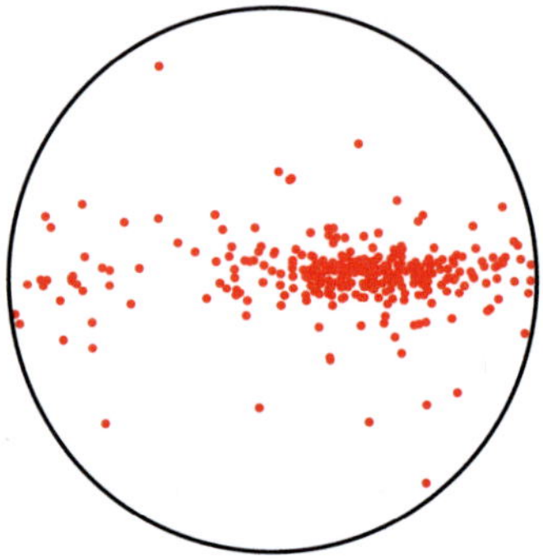

(a) Layer 8

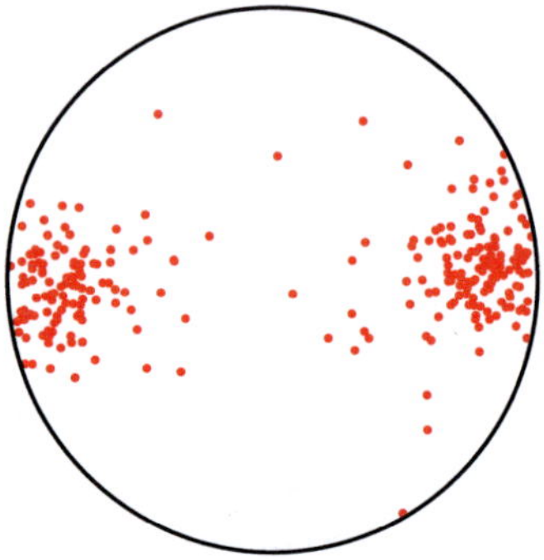

(b) Layer 11

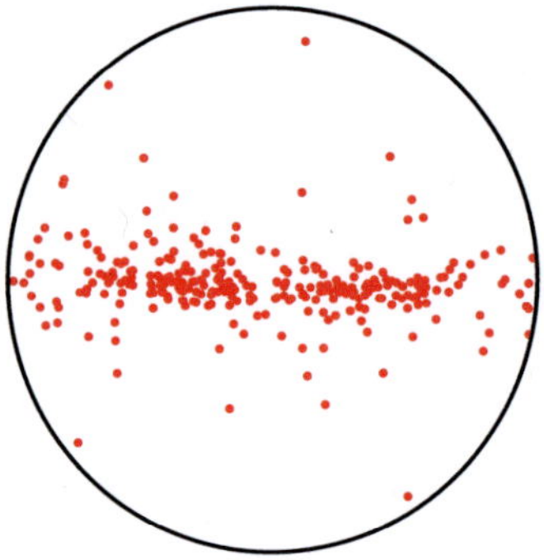

(c) Layer 14

Figure 3.10: Pole figures for the layers 8, 11 and 14 of μCT data at position (00)

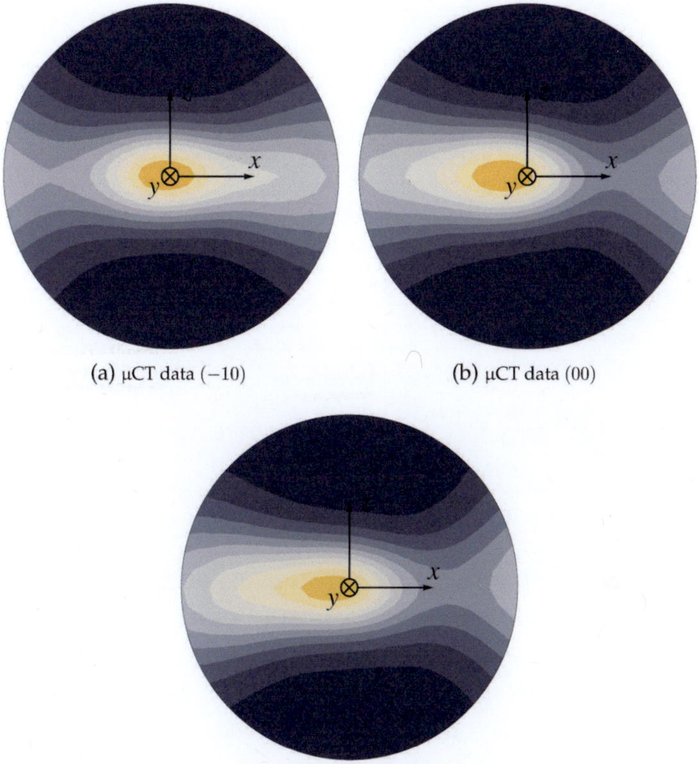

(a) μCT data (−10)

(b) μCT data (00)

(c) μCT data (10)

Figure 3.11: Stereographic projection of fiber axes for the μCT data (−10), (00) and (10)

Table 3.2: Number of fibers, mean length with standard deviation, diagonal components and eigenvalues of $\boldsymbol{N}$ for three µCT datasets

Data	(−10)
N	6329
$\bar{l}\,[\mu\mathrm{m}]$	330.28 ± 165.9
$\bar{N}_{11};\ \bar{N}_{22};\ \bar{N}_{33}$	$(0.381; 0.593; 0.026)$
$\mathrm{eig}(\bar{\boldsymbol{N}})$	$(0.594; 0.380; 0.026)$
Data	**(00)**
N	6355
$\bar{l}\,[\mu\mathrm{m}]$	330.18 ± 167.4
$\bar{N}_{11};\ \bar{N}_{22};\ \bar{N}_{33}$	$(0.392; 0.584; 0.024)$
$\mathrm{eig}(\bar{\boldsymbol{N}})$	$(0.635; 0.341; 0.024)$
Data	**(10)**
N	6200
$\bar{l}\,[\mu\mathrm{m}]$	332.66 ± 164.3
$\bar{N}_{11};\ \bar{N}_{22};\ \bar{N}_{33}$	$(0.404; 0.571; 0.025)$
$\mathrm{eig}(\bar{\boldsymbol{N}})$	$(0.613; 0.361; 0.025)$

wise constant properties. The microstructure of this composite consists of a matrix reinforced with N fibers.

The matrix phase is specified by the stiffness tensor $\mathbb{C}_{\mathsf{M}}$ and the volume fraction c_{M}. Each fiber is characterized with the stiffness $\mathbb{C}_\alpha$, the volume fraction c_α, the length l_α, the diameter d_α and the orientation of its axis $\boldsymbol{n}_\alpha$.
The fibers are approximated by spheroids of equal length and volume. Thus, the half-axes of the spheroid are correlated with the length and diameter of the corresponding fiber according to the following prescription: $2a_{1\alpha} = l_\alpha$ and $2a_{2\alpha} = \sqrt{3/2}\,d_\alpha$, whereby $a_{1\alpha}$ and $a_{2\alpha}$ are the first and the second half-axis of the spheroid related to fiber α.

The total fiber volume fraction is $c_{\mathsf{F}} = 1 - c_{\mathsf{M}} = \sum_{\alpha=1}^{N} c_\alpha$. The material behavior of the fibers is assumed to be uniform, i.g., $\mathbb{C}_\alpha = \mathbb{C}_{\mathsf{F}} \, \forall \alpha$.

3.4.2 Self-consistent homogenization

The effective elastic stiffness tensor $\bar{\mathbb{C}}$ can be formulated as ensemble average using the strain localization tensor $\mathbb{A}$ (Kröner, 1977).

$$\bar{\mathbb{C}} = \langle \mathbb{C}\mathbb{A} \rangle. \tag{3.5}$$

The strain localization tensor is symmetric and meets the normalization condition

$$\langle \mathbb{A} \rangle = \mathbb{I}^{\mathsf{s}}. \tag{3.6}$$

In case of ergodic media, the ensemble average can be interpreted as a volume average.

For microstructures with piecewise constant properties, the effective elastic stiffness is then given by

$$\bar{\mathbb{C}} = \mathbb{C}_{\mathsf{M}} + \sum_{\alpha=1}^{N} c_\alpha \left(\mathbb{C}_\alpha - \mathbb{C}_{\mathsf{M}} \right) \mathbb{A}_\alpha, \tag{3.7}$$

where $\mathbb{A}_\alpha$ describes the average strain localization tensor in terms of phase α or fiber α, respectively. For the formulation of the effective stiffness in equation (3.7), the normalization condition from equation (3.6) has been used. The SC homogenization scheme generally assumes that each fiber is embedded in an infinite homogeneous matrix with the properties of the effective material $\bar{\mathbb{C}} = \mathbb{C}^{\mathsf{SC}}$. The inclusion problem is solved based on the single inclusion formula of

Eshelby (Eshelby, 1957). Thus, for the special case of an ellipsoidal approximation of the cylindrical fibers, the strain localization $\mathbb{A}_\alpha$ is a function of the effective stiffness $\mathbb{C}^{\mathrm{SC}}$, the stiffness $\mathbb{C}_\alpha$ and the geometry of the ellipsoid $\boldsymbol{Z}_\alpha$, which contains the information about the fiber axis orientation:

$$\mathbb{A}_\alpha = \mathbb{A}(\mathbb{C}^{\mathrm{SC}}, \mathbb{C}_\alpha, \boldsymbol{Z}_\alpha) = \left(\mathbb{I}^{\mathrm{s}} + \mathbb{P}^{\mathrm{SC}}_\alpha \left(\mathbb{C}_\alpha - \mathbb{C}^{\mathrm{SC}}\right)\right)^{-1}. \tag{3.8}$$

Herein, $\mathbb{P}^{\mathrm{SC}}_\alpha$ denotes Hill's polarization tensor (see equation 4.34 in Willis (1981)):

$$\mathbb{P}^{\mathrm{SC}}_\alpha(\mathbb{C}^{\mathrm{SC}}, \boldsymbol{Z}_\alpha) = \frac{1}{4\pi \det(\boldsymbol{Z}_\alpha)} \int_S \mathbb{H}(\mathbb{C}^{\mathrm{SC}}, \boldsymbol{n}) \left(\boldsymbol{n} \cdot \left(\boldsymbol{Z}_\alpha^{-\mathsf{T}} \boldsymbol{Z}_\alpha^{-1} \boldsymbol{n}\right)\right)^{-3/2} \mathrm{d}S, \tag{3.9}$$

with $\mathbb{H}(\mathbb{C}^{\mathrm{SC}}, \boldsymbol{n}) = \mathbb{I}^{\mathrm{s}}(\boldsymbol{K}^{-1} \square (\boldsymbol{n} \otimes \boldsymbol{n}))\mathbb{I}^{\mathrm{s}}$ and $\boldsymbol{K} = \mathbb{C}^{\mathrm{SC}}[\![\boldsymbol{n} \otimes \boldsymbol{n}]\!]$. The polarization tensor $\mathbb{P}^{\mathrm{SC}}_\alpha$ depends on the stiffness $\mathbb{C}^{\mathrm{SC}}$ and the ellipsoidal geometry of the fibers $\boldsymbol{Z}_\alpha$:

$$\|\boldsymbol{Z}_\alpha \boldsymbol{x}\|^2 = \boldsymbol{x} \cdot \left(\boldsymbol{Z}_\alpha^{\mathsf{T}} \boldsymbol{Z}_\alpha \boldsymbol{x}\right) \leq 1. \tag{3.10}$$

$\boldsymbol{x}$ denotes a position vector in the three-dimensional space. The inverse eigenvalues of $\boldsymbol{Z}_\alpha$ correspond to the half axis of the ellipsoid with the number α. In equation (3.9), $\mathrm{d}S$ is a surface element of the unit sphere $S := \{\boldsymbol{n} \in \mathbb{R}^3 : \|\boldsymbol{n}\| = 1\}$, and $\det(\boldsymbol{Z}_\alpha)$ represents the determinant of $\boldsymbol{Z}_\alpha$.

The stiffness $\mathbb{C}^{\mathrm{SC}}$ can be computed by solving the following implicit equation resulting from equations (3.7) and (3.8) and the SC approximation $\bar{\mathbb{C}} = \mathbb{C}^{\mathrm{SC}}$ (Willis, 1986):

$$\mathbb{C}^{\mathrm{SC}} = \mathbb{C}_{\mathsf{M}} + \sum_{\alpha=1}^{N} c_\alpha \left(\mathbb{C}_\alpha - \mathbb{C}_{\mathsf{M}}\right) \left(\mathbb{I}^{\mathrm{s}} + \mathbb{P}^{\mathrm{SC}}_\alpha \left(\mathbb{C}_\alpha - \mathbb{C}^{\mathrm{SC}}\right)\right)^{-1}. \tag{3.11}$$

This equation for the unknown tensorial quantity $\mathbb{C}^{\mathrm{SC}}$ is solved numerically using a Newton-Raphson algorithm combined with a line search procedure. To determine Hill's polarization tensor $\mathbb{P}_{\alpha}^{\mathrm{SC}}$ numerically, a Gaussian quadrature over the unit sphere according to Mura (1987) is used.

3.4.3 Interaction direct derivative estimate

The interaction direct derivative (IDD) estimate, proposed by Zheng and Du (2001) is based on the three-phase model. In the three-phase model, the inclusions are embedded in a finite matrix material region, directly. This inclusion-matrix cell itself is embedded in the unbounded initially unknown effective medium. The difference between the three-phase model and IDD can be identified in the estimation of the stresses in the inclusions: In contrary to the three-phase model, Zheng and Du (2001) assume an unbounded medium with the properties of the matrix for the embedding of the inclusion in the matrix material. Du and Zheng (2002) have proven that the stresses in the inclusions are well approximated by this assumption with an error of second order of the inclusion volume fraction. The advantage of the IDD is its explicit structure, which is valid for multi-phase composites with different material symmetries and distributions.

The following prescription gives the IDD estimation of the properties of the effective medium:

$$\mathbb{C}^{\mathrm{IDD}} = \mathbb{C}_{\mathsf{M}} + \left(\mathbb{I}^{\mathrm{s}} - \sum_{\beta=1}^{N} c_{\beta} \left(\mathbb{C}_{\beta} - \mathbb{C}_{\mathsf{M}} \right) \mathbb{N}_{\beta} \mathbb{P}_{\beta}^{\mathrm{D}} \right)^{-1} \sum_{\alpha=1}^{N} c_{\alpha} \left(\mathbb{C}_{\alpha} - \mathbb{C}_{\mathsf{M}} \right) \mathbb{N}_{\alpha}, \tag{3.12}$$

with $\mathbb{N}_\alpha = (\mathbb{I}^s + \mathbb{P}_\alpha(\mathbb{C}_\alpha - \mathbb{C}_M))^{-1}$. Here, $\mathbb{P}_\alpha = \mathbb{P}(\mathbb{C}_M, \boldsymbol{Z}_\alpha)$ is Hill's polarization tensor as defined in equation (3.9) but $\mathbb{C}^{SC}$ replaced with $\mathbb{C}_M$. If the matrix-inclusion cell takes on an ellipsoidal shape, then, $\mathbb{P}_\alpha^D = \mathbb{P}(\mathbb{C}_M, \boldsymbol{Z}_\alpha^D)$ is the polarization tensor corresponding to an ellipsoidal inclusion with geometry of the matrix-inclusion cell $\boldsymbol{Z}_\alpha^D$, which is embedded in an infinite matrix with the stiffness $\mathbb{C}_M$. The shape of the matrix-inclusion cell defines the inclusion distribution in the composite (Zheng and Du, 2001). In the present work, this shape is assumed to be equal to the shape of the spheroidal approximation $\boldsymbol{Z}_\alpha$ of the corresponding inclusion itself.

Particularly if the distribution is constant for all inclusions ($\boldsymbol{Z}_\alpha^D = \boldsymbol{Z}^D$), the IDD estimate is equivalent to the Hashin-Shtrikman estimate of Ponte Castañeda and Willis (1995), which is based on the Hashin-Shtrikman variational structure in the form developed by Willis (1977, 1978). A detailed discussion of the relation of IDD to the estimate of Ponte Castañeda and Willis (1995) and, e.g., Mori-Tanaka can be found in Zheng and Du (2001) and Du and Zheng (2002).

3.4.4 A two-step bounding method

The SC and IDD approaches deliver *estimates* of the effective elastic properties. The two-step *bounding* method (TS) is a simple method providing an admissible range of possible effective properties. It should be noted, however, that the TS homogenization applying the HS method based on constant stress polarization and two subsequent steps does not represent a rigorous second-order bounding scheme.

The first step of the TS approach consists of a decomposition of the microstructure into as many domains as there are different fibers (Pierard et al., 2004). Thus, fibers of the same shape and orientation are aggregated into unidirectional domains. The fiber volume fraction in the domains corresponds to the total fiber volume fraction c_{F}. For each of these domains, the upper and lower bounds, denoted by $\mathbb{C}_{\alpha}^{\mathrm{UD}+}$ and $\mathbb{C}_{\alpha}^{\mathrm{UD}-}$, respectively, are calculated using the unidirectional (UD) special case of the second-order HS bounds. Assuming the fiber material being stiffer than the matrix material leads to the following prescriptions for $\mathbb{C}_{\alpha}^{\mathrm{UD}+}$ and $\mathbb{C}_{\alpha}^{\mathrm{UD}-}$ (Willis, 1977):

$$\begin{aligned} \mathbb{C}_{\alpha}^{\mathrm{UD}+} &= \mathbb{C}_{\alpha} + (1 - c_{\mathsf{F}})\left(\mathbb{C}_{\mathsf{M}} - \mathbb{C}_{\alpha}\right)\left(\mathbb{I}^{\mathrm{s}} + c_{\mathsf{F}}\mathbb{P}_{\alpha}^{\mathrm{UD}}\left(\mathbb{C}_{\mathsf{M}} - \mathbb{C}_{\alpha}\right)\right)^{-1}, \\ \mathbb{C}_{\alpha}^{\mathrm{UD}-} &= \mathbb{C}_{\mathsf{M}} + c_{\mathsf{F}}\left(\mathbb{C}_{\alpha} - \mathbb{C}_{\mathsf{M}}\right)\left(\mathbb{I}^{\mathrm{s}} + (1 - c_{\mathsf{F}})\mathbb{P}_{\alpha}^{\mathrm{UD}}\left(\mathbb{C}_{\alpha} - \mathbb{C}_{\mathsf{M}}\right)\right)^{-1}. \end{aligned} \tag{3.13}$$

Hill's polarization tensor $\mathbb{P}_{\alpha}^{\mathrm{UD}}$ is known explicitly for the case of unidirectionally aligned ellipsoidal inclusions. The resulting bounds exhibit a transversely isotropic linear elastic behavior. If each fiber is unique in terms of direction, geometry or material properties, then N domains have to be considered. In equation (3.13), α depicts the number of the particular domain with $\alpha \in \{1, \ldots, N\}$.

Within the second step, again HS bounds are calculated assuming an isotropic two-point correlation function for the domains (Ponte Castañeda and Suquet, 1997). Herein, only corresponding bounds are combined: the lower (upper) bound of the domains is homogenized with the lower (upper) bound for the granular structure. The resulting stiffness tensors are denoted by $\mathbb{C}^{\mathrm{TS-}}$ and $\mathbb{C}^{\mathrm{TS+}}$, respectively:

$$\mathbb{C}^{\mathrm{TS}\pm} = \sum_{\alpha=1}^{N} \frac{c_{\alpha}}{c_{\mathsf{F}}} \mathbb{C}_{\alpha}^{\mathrm{UD}\pm} \mathbb{A}_{\alpha}^{\pm} = \sum_{\alpha=1}^{N} \frac{c_{\alpha}}{c_{\mathsf{F}}} \mathbb{C}_{\alpha}^{\mathrm{UD}\pm} \mathbb{M}_{\alpha}^{\pm} \left\langle \mathbb{M}^{\pm} \right\rangle^{-1}, \tag{3.14}$$

with

$$\mathbb{M}_\alpha^\pm = \left(\mathbb{I}^s + \mathbb{P}_0(\mathbb{C}_\alpha^{\mathrm{UD}\pm} - \mathbb{C}_0^\pm)\right)^{-1},$$
$$\langle \mathbb{M}^\pm \rangle = \sum_{\beta=1}^{N} \frac{c_\beta}{c_{\mathrm{F}}} \left(\mathbb{I}^s + \mathbb{P}_0(\mathbb{C}_\beta^{\mathrm{UD}\pm} - \mathbb{C}_0^\pm)\right)^{-1}. \tag{3.15}$$

In the last equation, the definition of $\mathbb{P}_0$ depends on modeling assumptions concerning the shape and distribution of the domains. Assuming an isotropic two-point correlation of the domains, $\mathbb{P}_0$ corresponds to the spherical polarization tensor. Another possible assumption would be an ellipsoidal two-point correlation. Due to simplicity, the isotropic two-point correlation has been assumed. In the case of the upper (lower) HS bound, for $\mathbb{C}_0^\pm$ the maximum (minimum) isotropic part of all stiffness tensors of the domains $\mathbb{C}_\alpha^{\mathrm{UD}+}$ ($\mathbb{C}_\alpha^{\mathrm{UD}-}$) is taken (Nadeau and Ferrari, 2001).

3.5 Results and discussion

The mean values of Young's moduli and Poisson's ratios of the isotropic polypropylene material in the strain region 0.0005-0.0025, as shown in Tab. 3.1, were used as input parameters for the homogenization of the composite. The fiber mass fraction is $30\text{wt.}\%$, what corresponds to a fiber volume fraction of $c_{\mathrm{F}} = 0.13$.

Combining this data with the segmented µCT data, the effective elastic properties were calculated according to the methods described in the foregoing section. To compare the experimentally measured Young's moduli to the computational results, for each homogenized stiffness, the direction-dependent Young's modulus(Böhlke and Brüggemann, 2001) $E(\boldsymbol{d})$, given in equation (3.16), was determined and evaluated in the appropriate direction:

$$\frac{1}{E(\boldsymbol{d})} = \boldsymbol{d} \otimes \boldsymbol{d} \cdot \mathbb{S}\left[\boldsymbol{d} \otimes \boldsymbol{d}\right]. \tag{3.16}$$

In combination with the direction-dependent bulk modulus $K(\boldsymbol{d})$

$$\frac{1}{3K(\boldsymbol{d})} = \boldsymbol{I} \cdot \mathbb{S}\left[\boldsymbol{d} \otimes \boldsymbol{d}\right] \tag{3.17}$$

the linear elastic behaviour is completely defined. In equation (3.16), the direction $\boldsymbol{d}$ is parametrized with spherical coordinates, and $\mathbb{S} = \mathbb{C}^{-1}$ is the compliance tensor.

In the following, first, the numerical results based on the three segmented µCT datasets are compared to each other, and second, the numerical estimations are compared to the experimentally measured Young's moduli. In Figure 3.12, the contours of the direction-dependent Young's moduli in the x-y plane are shown in dependence of the azimuth angle as depicted in Figure 2.2. It can be seen that the SC and IDD estimates are located between the bounding TS results for all three datasets. SC predicts a stiffer behavior than IDD. The direction-dependence of Young's modulus is more pronounced by SC and IDD than by the TS methods. The TS bounding methods exhibit a rather large range of admissible Young's moduli. Furthermore, in Figure 3.12, it can be observed that the direction-dependence of Young's moduli is very similar for the (00) and the (10) data, see Figure 3.12(b) and Figure 3.12(c). The data at the position (-10) yields a different direction-dependence of the Young's modulus, see Figure 3.12(a). This can be attributed to the flow conditions during the manufacturing process. The same phenomenon is visible in the stereographic projections of all fibers in Figure 3.11. The maximum Young's modulus is located apart from the filling di-

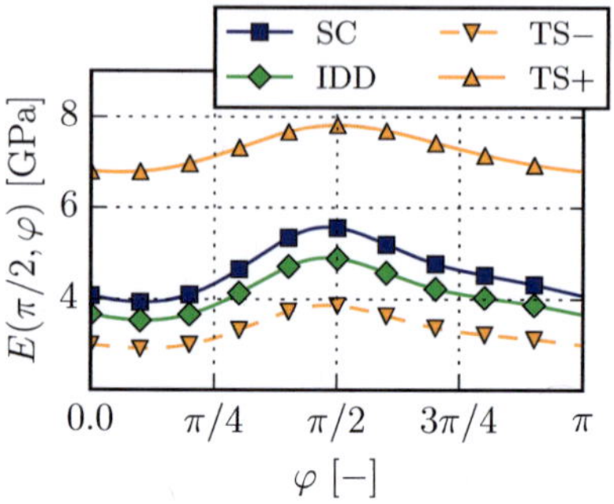

(a) μCT (-10)

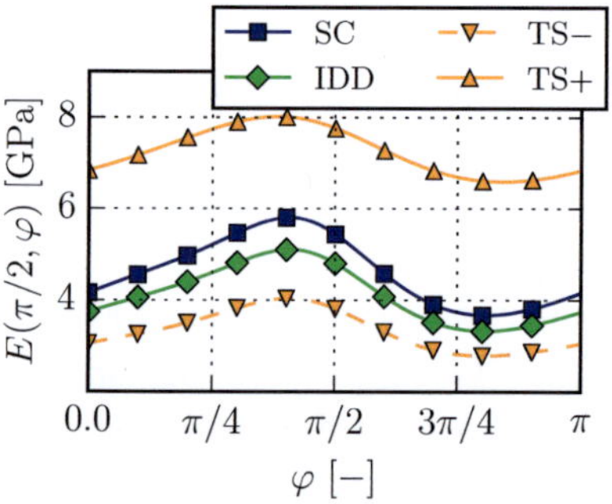

(b) μCT (00)

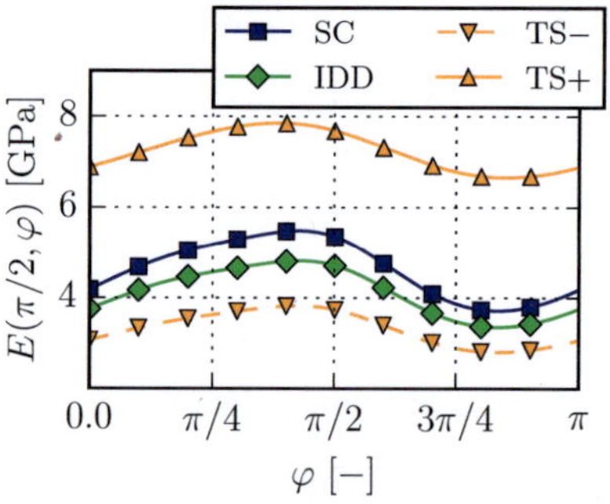

(c) μCT (10)

Figure 3.12: Comparison of numerical Young's moduli for the three segmented μCT datasets (see Figure 3.6)

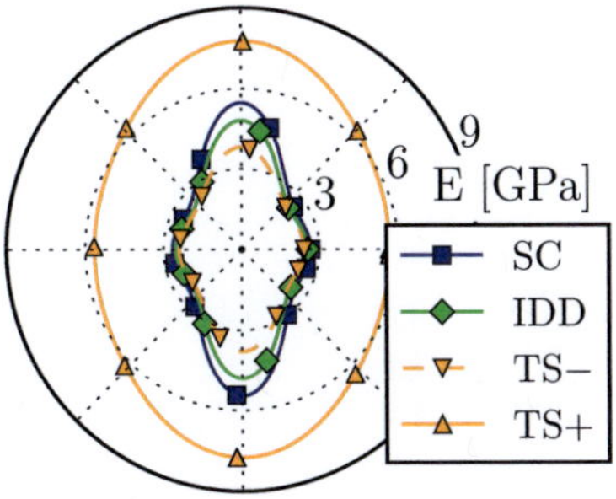

(a) x_1-x_3 plane

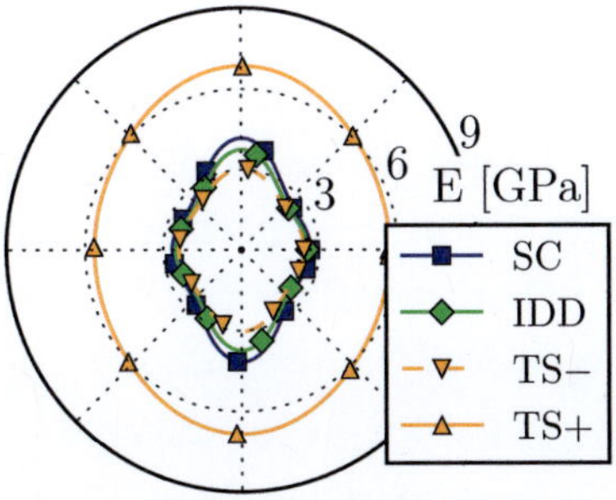

(b) x_2-x_3 plane

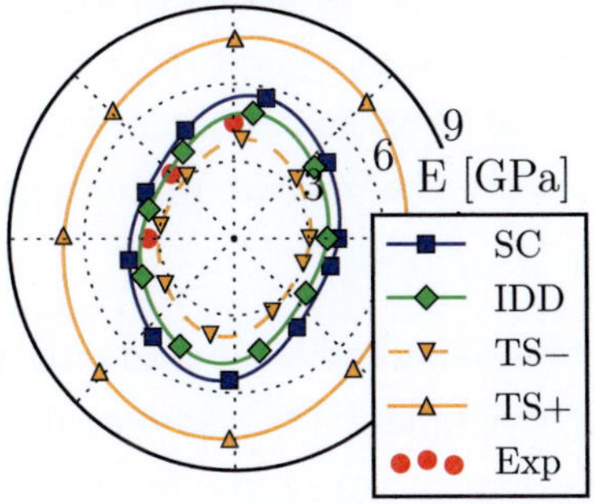

(c) x_1-x_2 plane

Figure 3.13: Orientation dependence of Young's modulus in the (a) x_1-x_3 plane, (b) x_2-x_3 plane, (c) x_1-x_2 plane for the overall data

Table 3.3: Comparison of experimental and numerical Young's modulus E [GPa] for the overall dataset

	E^{exp}	$E^{\text{TS-}}$	E^{IDD}	E^{SC}	$E^{\text{TS+}}$
0°	4.482	3.798	4.809	5.463	7.754
90°	3.452	3.035	3.741	4.167	6.842
45°	3.541	2.986	3.661	4.087	6.912

rection ($\boldsymbol{e}_2$-direction), which is also an effect of the inhomogeneous flow process.

Since the volume element analyzed with the μCT measurements is small compared to the size of the tensile specimen, the three μCT datasets are combined to an overall dataset for the comparison of the numerical and experimental results. In Figure 3.13, the shapes of the direction-dependent Young's moduli are shown on the x_1-x_3, x_2-x_3, and x_1-x_2 plane for the overall data. As the specimens for the tensile test were prepared in the x_1-x_2 plane, the experimental results are added to Figure 3.13(c). It can be observed, that the direction-dependent Young's modulus of the overall data exhibits a defined symmetry in the x_1-x_3 and the x_2-x_3 plane. In the x_1-x_2 plane, no characteristic symmetry is obvious. In Tab. 3.3, the experimental and the numerical results for all homogenization methods for the overall data are given.

In addition to the qualitative evaluation of the numerical results compared to experimental measurements in Figure 3.13 and Tab. 3.3, it is interesting to notice that the predicted anisotropy of the direction-dependent Young's modulus essentially depends on the homogenization procedure. Especially the upper bound of the two-step approach does not represent the anisotropy accurately. It predicts a rather isotropic material behavior.

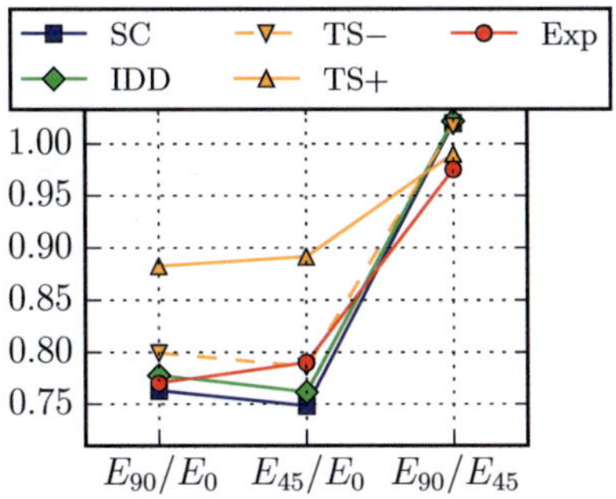

Figure 3.14: Anisotropy ratios of Young's modulus for the overall μCT data

To observe this more quantitatively, the ratios E_{90}/E_0, E_{45}/E_0 and E_{90}/E_{45} between the Young's modulus values in different directions were calculated for experimental and numerical results separately. In Figure 3.14, the comparison of these ratios is given for the overall μCT data.

The TS methods predict a smaller anisotropy for E_{90}/E_0 and E_{90}/E_{45} and TS- also for E_{45}/E_0. SC predicts a greater anisotropy for E_{90}/E_0 and E_{45}/E_0 and IDD only for E_{45}/E_0. The lower bound of TS reflects the experimental anisotropy ratios better than the upper bound. Generally, all ratios of the upper bound are on a similar elevated level. This conveys a nearly isotropic material behavior.

3.6 Summary and conclusions

This chapter discusses the examination of the thermoplastic composite PPGF30 through tensile tests in three different directions, and its microstructure through μCT measurements in combination with a stochastic fiber analysis approach. The microstructure was analyzed by means of the resulting segmented µCT data. Three different mean-field homogenization methods were applied to estimate the effective elastic properties directly based on segmented microstructural data.

Relying on the experimental and numerical results, the following conclusions can be made:

- **Anisotropy**: The evaluated injection-molded SFRC specimens show an anisotropic material behavior. The experiments in different material directions, indicate the importance of attaining detailed knowledge of the microstructure in order to dimension parts made of such a material.

- **Inhomogeneity**: Important properties of the microstructure, like the fiber orientation distribution and fiber length distribution, can be analyzed by using the segmented µCT data. Thus, a layer-wise analysis reveals the dependence of these properties on the position in the specimen. Thereby, a more accurate observation of the microstructure of SFRCs is possible.

- **Microstructural information**: All presented mean-field homogenization approaches consider the segmented µCT data and take advantage of the information about the distribution of the fiber axes, fiber lengths and fiber radii.

- **Two-step homogenization**: In order to get an admissible range for the effective properties, two-step bounding methods have been applied: In the first step, the Hashin-Shtrikman bounds for unidirectional aligned fiber domains and, in the second step, the Hashin-Shtrikman bounds for granular systems have been calculated.

- **SC and IDD estimates**: Both methods, SC and IDD, allow for a direct consideration of the segmented μCT data. IDD uses the matrix material as reference material. SC is an implicit approach using the effective properties as reference material. In addition to the shape, the orientation and the material properties, IDD accounts for a statistical description of the distribution of the inclusions. Here, an spheroidal distribution equal to the shape of the spheroidal approximation of the fibers have been used.

- **Experimental vs numerical results**: The self-consistent homogenization method delivers a stiffer material behavior compared to experimental measurements. Due to the inherent characteristics of the SC approach (Hill, 1965; Torquato, 2002), this method tends to overestimate the elastic properties for the case of hard inclusions embedded in a soft matrix similar to the problem at hand.

  The Young's modulus values predicted by IDD are located between the lower TS bounds and SC. Like SC, this approach delivers also a stiffer material behavior compared to experiments. The IDD predictions of the Young's modulus values show the closest agreement with experimental measurements.

  In terms of the anisotropy ratios, SC and IDD show similar trends. Both methods predict a close agreement with experiments for E_{90}/E_0, a larger anisotropy for E_{45}/E_0 and a smaller for E_{90}/E_{45}.

  The TS methods, except TS- for E_{45}/E_0, show larger anisotropy-ratios compared to experiments. Two reasons are responsible for that: First, in the second step, the reference material has to be isotropic. Therefore, the minimal and maximal isotropic part of the domain stiffnesses, respectively, has been taken as reference material. Second, an isotropic two-point correlation function has been assumed by applying the spherical polarization tensor for TS- and TS+.

The present investigation shows the modeling of the elastic behavior of SFRCs, using homogenization techniques in combination with detailed microstructural data from a micro-computed tomography image analysis method. The homogenization results demonstrate the need of realistic microstructural data for the simulation of parts made of SFRCs. Relying on the segmented μCT data, the IDD approximation of the elastic properties delivers the closest agreement with experimental results for the considered composite PPGF30. PPGF30 is, however, a thermoplastic based composite. Thus, the predictions of the effective material behavior can be improved by taking viscous effects and defects especially at the fiber ends into account.

4 Mean-field vs full-field voxel based homogenization methods[1]

4.1 Introduction

In this chapter, the mean-field homogenization methods described in section 3.4 are opposed to a voxel-based full-field approach. For this purpose, artificially generated microstructures with unidirectionally aligned, and two configurations with misaligned fibers are considered, as well. The misaligned configurations are inspired by the characteristics of the microstructure of injection molded SFRCs. In addition, for all microstructures the phase contrast of the elastic moduli is varied. Phase contrast of 44, 100, and 1000 are considered. The phase contrast of 44 corresponds to the relation of Young's moduli of glass and polypropylene. For the unidirectional configuration the volume fractions of fibers of 13%, 17% and 21% are taken into account. With this variation of parameters, it is possible to reveal trends in the prediction of the elastic properties of the methods applied, and to draw conclusions also on other cases.

[1] This chapter is based on the paper "Homogenization of linear elastic properties of short-fiber reinforced composites – A comparison of mean-field and voxel-based methods" (Müller et al., 2015b)

4.2 Determination of elastic properties

4.2.1 Modeling preliminaries

Equivalently to the forgoing chapter, again SFRCs are considered, where the matrix and the arbitrary oriented fibers are assumed to be linear isotropic with piecewise constant properties. The matrix is characterized by its stiffness tensor $\mathbb{C}_{\mathsf{M}}$ and volume fraction c_{M}. Accordingly, the fiber with number α is specified by the orientation of the axis $\boldsymbol{n}_\alpha$, the length l_α, the radius r_α, the volume fraction c_α and the stiffness tensor $\mathbb{C}_\alpha$. The total fiber volume fraction is indicated with $c_{\mathsf{F}} = 1 - c_{\mathsf{M}} = \sum_{\alpha=1}^{N} c_\alpha$.

4.2.2 Full-field homogenization[2]

Periodic boundary value problem

In contrary to the mean-field methods presented in section 3.4, for the numerical homogenization using full-field approaches, the microstructure is fully resolved within a representative volume element (RVE). The RVE is often of rectangular shape and it is also referred to as unit-cell or statistic volume element. On this RVE, which represents a heterogeneous periodic medium with position dependent stiffness $\mathbb{C}(\boldsymbol{x})$, a periodic boundary value problem for the displacement fluctuations $\tilde{\boldsymbol{u}}$ has to

[2] The theory presented in this chapter, and the appropriate implementation have been elaborated by M. Kabel and H. Andrä (Müller et al., 2015b). For the sake of completeness, the theory is reproduced in this chapter.

be solved in order to obtain the homogeneous material behavior:

$$\begin{aligned} \operatorname{div}(\boldsymbol{\sigma}(\boldsymbol{x})) &= \mathbf{0}, & \boldsymbol{x} &\in \Omega, \\ \boldsymbol{\sigma}(\boldsymbol{x}) &= \mathbb{C}(\boldsymbol{x})[\boldsymbol{\varepsilon}(\boldsymbol{x})], & \boldsymbol{x} &\in \Omega, \\ \boldsymbol{\varepsilon}(\boldsymbol{x}) &= \bar{\boldsymbol{\varepsilon}} + \operatorname{sym}(\operatorname{grad}(\tilde{\boldsymbol{u}}(\boldsymbol{x}))), & \boldsymbol{x} &\in \Omega, \\ \tilde{\boldsymbol{u}}(\boldsymbol{x}) &\quad \text{periodic}, & \boldsymbol{x} &\in \partial\Omega, \\ (\boldsymbol{\sigma}\boldsymbol{n})(\boldsymbol{x}) &\quad \text{anti}-\text{periodic}, & \boldsymbol{x} &\in \partial\Omega. \end{aligned} \tag{4.1}$$

This periodic boundary value problem is uniquely solvable for all constant strains $\bar{\boldsymbol{\varepsilon}}$ in the space

$$\tilde{\boldsymbol{u}} \in \left(H^{1,\#}(\Omega)\right)^3 : \langle \tilde{\boldsymbol{u}} \rangle_\Omega := \int_\Omega \tilde{\boldsymbol{u}}(x)\, \mathrm{d}\Omega = \mathbf{0}, \tag{4.2}$$

where $H^{1,\#}(\Omega)$ is the closure of $C^{\infty,\#}(\Omega) = \{\boldsymbol{v} \in C^\infty(\Omega) : \boldsymbol{v} \text{ periodic}\}$ in $H^1(\Omega)$ (see, e.g., Bakhvalov and Panasenko (1989); Cioranescu and Donato (1999)). In this work, the microstructure is represented in a binary manner using a regular three-dimensional voxel discretization within the RVE.

Equivalent strain and stress based Lippmann-Schwinger equation

The constitutive equation for the stress $\boldsymbol{\sigma}(\boldsymbol{x}) = \mathbb{C}(\boldsymbol{x})[\boldsymbol{\varepsilon}(\boldsymbol{x})]$ can be rewritten into

$$\boldsymbol{\sigma}(\boldsymbol{x}) = \mathbb{C}_0[\boldsymbol{\varepsilon}(\boldsymbol{x})] + \boldsymbol{p}(\boldsymbol{x}). \tag{4.3}$$

In the foregoing equation, $\mathbb{C}_0$ is the stiffness tensor of a homogeneous comparison material and $\boldsymbol{p}(\boldsymbol{x})$ the stress polarization, which is defined as

$$\boldsymbol{p}(\boldsymbol{x}) = (\mathbb{C}(\boldsymbol{x}) - \mathbb{C}_0)\,[\boldsymbol{\varepsilon}(\boldsymbol{x})]. \tag{4.4}$$

If the stress polarization $\boldsymbol{p}(\boldsymbol{x})$ with support in Ω, e.g., $\operatorname{supp}(\boldsymbol{p}) \subseteq \Omega$ is known, the solution of $\operatorname{div}(\boldsymbol{\sigma}(\boldsymbol{x})) = \mathbf{0}$, see equation (4.1), can be depicted as (Kröner, 1977)

$$\boldsymbol{\varepsilon}(\boldsymbol{x}) = \bar{\boldsymbol{\varepsilon}} - (\mathbb{G}_0 * \boldsymbol{p})(\boldsymbol{x}), \tag{4.5}$$

where $\mathbb{G}_0$ is the nonlocal elastic Green operator for strains associated with the comparison material $\mathbb{C}_0$. The convolution in equation (4.5) is defined by

$$(\mathbb{G}_0 * \boldsymbol{p})(\boldsymbol{x}) = \int_\Omega \mathbb{G}_0(\boldsymbol{x} - \boldsymbol{y})[\boldsymbol{p}(\boldsymbol{y})] \, \mathrm{d}\Omega. \tag{4.6}$$

Hence, the strain based formulation of the Lippmann-Schwinger equation can be derived by considering equation (4.4) in equation (4.5)

$$(\mathbb{I}^{\mathrm{s}} + \mathbb{B}_{\boldsymbol{\varepsilon}})[\boldsymbol{\varepsilon}(\boldsymbol{x})] := \boldsymbol{\varepsilon}(\boldsymbol{x}) + (\mathbb{G}_0 * ((\mathbb{C} - \mathbb{C}_0)[\boldsymbol{\varepsilon}(\boldsymbol{x})]))(\boldsymbol{x}) = \bar{\boldsymbol{\varepsilon}}. \tag{4.7}$$

This integral equation for the strain is equivalent to an integral equation for the stress (Kröner, 1971)

$$(\mathbb{I}^{\mathrm{s}} + \mathbb{B}_{\boldsymbol{\sigma}})[\boldsymbol{\sigma}(\boldsymbol{x})] := \boldsymbol{\sigma}(\boldsymbol{x}) + (\mathbb{L}_0 * ((\mathbb{S} - \mathbb{S}_0)[\boldsymbol{\sigma}(\boldsymbol{x})]))(\boldsymbol{x}) = \bar{\boldsymbol{\sigma}}, \tag{4.8}$$

with the local compliance $\mathbb{S}(\boldsymbol{x}) = \mathbb{C}^{-1}(\boldsymbol{x})$, the compliance of the comparison material $\mathbb{S}_0 = \mathbb{C}_0^{-1}$, the macroscopic stress $\bar{\boldsymbol{\sigma}} = \mathbb{C}_0[\bar{\boldsymbol{\varepsilon}}]$ and the Green operator $\mathbb{L}_0$ for stresses associated with the comparison material. Using the following prescription

$$\mathbb{L}_0 = \mathbb{C}_0 - \mathbb{C}_0\mathbb{G}_0\mathbb{C}_0, \tag{4.9}$$

the Green operator $\mathbb{L}_0$ for stresses can be determined in dependence of the Green operator $\mathbb{G}_0$ for strains. For the sake of a simpler notation, the dependency of the quantities on the spatial position $\boldsymbol{x}$ is not noted in the following equations anymore.

Energy and strain equivalence principle

Taking Hill's condition, also known as the energy equivalence principle,

$$\frac{1}{2}\langle \boldsymbol{\varepsilon}_\alpha \cdot \mathbb{C}[\boldsymbol{\varepsilon}_\beta]\rangle = \frac{1}{2}\langle \boldsymbol{\varepsilon}_\alpha\rangle \cdot \bar{\mathbb{C}}[\langle \boldsymbol{\varepsilon}_\beta\rangle] \tag{4.10}$$

into consideration, the effective and homogeneous stiffness $\bar{\mathbb{C}}$ can be determined. For this purpose, the boundary value problem in equation (4.1) has to be solved for a basis of the six dimensional space of macroscopic strains $\bar{\boldsymbol{\varepsilon}}$. Hence, $\boldsymbol{\varepsilon}_\alpha$ or $\boldsymbol{\varepsilon}_\beta$, respectively, denote the six necessary linear independent loading cases with $\alpha, \beta \in \{1, \ldots, 6\}$. The same can be done for the effective compliance $\bar{\mathbb{S}}$

$$\frac{1}{2}\langle \boldsymbol{\sigma}_\alpha \cdot \mathbb{S}[\boldsymbol{\sigma}_\beta]\rangle = \frac{1}{2}\langle \boldsymbol{\sigma}_\alpha\rangle \cdot \bar{\mathbb{S}}[\langle \boldsymbol{\sigma}_\beta\rangle]. \tag{4.11}$$

The effective stiffness and compliance tensors can then be calculated by (Bishop and Hill, 1951b,a)

$$\langle \mathbb{C}[\boldsymbol{\varepsilon}_\beta]\rangle = \bar{\mathbb{C}}[\langle \boldsymbol{\varepsilon}_\beta\rangle], \tag{4.12}$$

$$\langle \mathbb{S}[\boldsymbol{\sigma}_\beta]\rangle = \bar{\mathbb{S}}[\langle \boldsymbol{\sigma}_\beta\rangle]. \tag{4.13}$$

Lippmann-Schwinger equations for the Hashin-Shtrikman bounds

In the case of $\mathbb{C} \leq \mathbb{C}_0$ ($\mathbb{C} \geq \mathbb{C}_0$), the energy principle of Hashin and Shtrikman (1962b) states that the following bounds on the elastic energy in the unit-cell hold true for an arbitrary stress polarization field $\boldsymbol{p}(\boldsymbol{x})$:

$$\langle \boldsymbol{\varepsilon} \cdot \mathbb{C}[\boldsymbol{\varepsilon}] \rangle \underset{(\geq)}{\leq} \bar{\boldsymbol{\varepsilon}} \cdot \mathbb{C}_0[\bar{\boldsymbol{\varepsilon}}] + 2 \langle \boldsymbol{p} \rangle \cdot \bar{\boldsymbol{\varepsilon}} - \left\langle \boldsymbol{p} \cdot (\mathbb{C} - \mathbb{C}_0)^{-1} [\boldsymbol{p}] \right\rangle \\ - \langle \boldsymbol{p} \cdot (\mathbb{G}_0 * \boldsymbol{p}) \rangle . \tag{4.14}$$

If the stress polarization takes on the form given in equation (4.4), equality is obtained in the last equation.

Willis (1977) has shown that the Hashin-Shtrikman bounds are quadratic positive (negative) definite forms on the space of all polarization fields. Thus, an optimal $\boldsymbol{p}^{\mathsf{HS}}$ has to exist in the subspace of voxel-wise constant polarization fields. The bounds in equation (4.14) are applied exactly to this subspace. By introducing the periodized Green operator $\mathbb{G}_0^{\#}$, the bounds (4.14) or $\mathbb{G}_0 * \boldsymbol{p}$, respectively, can be calculated by a discrete fast Fourier transformation in the subspace of voxel-wise constant polarization fields without any approximation error.

The application of this result for the polarization fields to the associated strain field

$$\boldsymbol{\varepsilon}^{\mathsf{HS}} := (\mathbb{C} - \mathbb{C}_0)^{-1} [\boldsymbol{p}^{\mathsf{HS}}], \tag{4.15}$$

results in

$$\boldsymbol{\varepsilon}^{\mathsf{HS}} + \mathbb{G}_0^{\#} * \left((\mathbb{C} - \mathbb{C}_0) [\boldsymbol{\varepsilon}^{\mathsf{HS}}] \right) = \bar{\boldsymbol{\varepsilon}}. \tag{4.16}$$

This equation is of the same form as (4.7), what implies that the bounds of Hashin-Shtrikman can be calculated by the methods discussed for the strain based formulation of the Lippmann-Schwinger equation. The difficulties arising in the calculation of the periodized Green operator $\mathbb{G}_0^{\#}$ are discussed in (Brisard and Dormieux, 2010).

The stress based form of the Lippmann-Schwinger equation can be rewritten, accordingly. With this formulation, rigid inclusions can be treated without problems. Taking the eigenstrain $\varepsilon^\star$ according to Hill (1963) into consideration

$$\varepsilon^\star = \mathbb{S}_0[\boldsymbol{p}] = (\mathbb{S}_0 - \mathbb{S})\,[\boldsymbol{\sigma}], \tag{4.17}$$

the bounds of Hashin-Shtrikman can be formulated as follows (Willis, 1977):

$$\begin{aligned} \langle \boldsymbol{\sigma} \cdot \mathbb{S}[\boldsymbol{\sigma}] \rangle \underset{(\geq)}{\leq} \bar{\boldsymbol{\sigma}} \cdot \mathbb{S}_0[\bar{\boldsymbol{\sigma}}] + 2\,\langle \varepsilon^\star \rangle \cdot \bar{\boldsymbol{\sigma}} + \left\langle \varepsilon^\star \cdot (\mathbb{S} - \mathbb{S}_0)^{-1}\,[\varepsilon^\star] \right\rangle \\ + \langle \varepsilon^\star \cdot (\mathbb{L}_0 * \varepsilon^\star) \rangle\,, \end{aligned} \tag{4.18}$$

if $\mathbb{C} \leq \mathbb{C}_0$ ($\mathbb{C} \geq \mathbb{C}_0$) or equivalently $\mathbb{S} \geq \mathbb{S}_0$ ($\mathbb{S} \leq \mathbb{S}_0$). Equivalently to equation (4.14), equality is reached for $\varepsilon^\star = (\mathbb{S}_0 - \mathbb{S})[\boldsymbol{\sigma}]$. Applying the above arguments to the associated stress field $\boldsymbol{\sigma}^{\mathsf{HS}} := (\mathbb{S}_0 - \mathbb{S})^{-1}\,[\varepsilon^{\star\mathsf{HS}}]$ of the optimal eigenstrain field $\varepsilon^{\star\mathsf{HS}}$ in the subspace of voxel-wise constant eigenstrain fields, the following equation is obtained

$$\boldsymbol{\sigma}^{\mathsf{HS}}(\boldsymbol{x}) + \left(\mathbb{L}_0^{\#} * \left((\mathbb{S} - \mathbb{S}_0)\,[\boldsymbol{\sigma}^{\mathsf{HS}}]\right)\right)(\boldsymbol{x}) = \bar{\boldsymbol{\sigma}} \tag{4.19}$$

with $\mathbb{L}_0^{\#}$ being the periodized Green operator for stresses. This operator is defined equivalent to the periodized Green operator for strains.

Numerical algorithm

The strain based (4.7) as well as the stress based formulation (4.8) of the Lippmann-Schwinger equation can be solved iteratively. Using the Neumann series expansion for the inversion of $\mathbb{I}^s + \mathbb{B}_{\varepsilon}$ or

$\mathbb{I}^s + \mathbb{B}_{\boldsymbol{\sigma}}$, the iterates of the local strain or stress, respectively, reads

$$\boldsymbol{\varepsilon}^n = \sum_{\alpha=0}^{n} (-\mathbb{B}_{\boldsymbol{\varepsilon}})^\alpha [\bar{\boldsymbol{\varepsilon}}], \qquad \boldsymbol{\sigma}^n = \sum_{\alpha=0}^{n} (-\mathbb{B}_{\boldsymbol{\sigma}})^\alpha [\bar{\boldsymbol{\sigma}}], \tag{4.20}$$

which can also be written as

$$\boldsymbol{\varepsilon}^0 = \bar{\boldsymbol{\varepsilon}}, \qquad \boldsymbol{\sigma}^0 = \bar{\boldsymbol{\sigma}}, \tag{4.21}$$

$$\boldsymbol{\varepsilon}^{n+1} = -\mathbb{B}_{\boldsymbol{\varepsilon}}[\boldsymbol{\varepsilon}^n] + \bar{\boldsymbol{\varepsilon}}, \qquad \boldsymbol{\sigma}^{n+1} = -\mathbb{B}_{\boldsymbol{\sigma}}[\boldsymbol{\sigma}^n] + \bar{\boldsymbol{\sigma}}. \tag{4.22}$$

The iterates in equation (4.22) can be efficiently computed in four steps using discrete fast Fourier transformation (DFT) by the so called basic scheme (BS) for polarization and eigenstrain, respectively (see Moulinec and Suquet (1994)):

$$\boldsymbol{p} = (\mathbb{C} - \mathbb{C}_0)[\boldsymbol{\varepsilon}^n], \qquad \boldsymbol{\varepsilon}^\star = (\mathbb{S} - \mathbb{S}_0)[\boldsymbol{\sigma}^n], \tag{4.23}$$

$$\hat{\boldsymbol{p}} = \mathrm{DFT}(\boldsymbol{p}), \qquad \hat{\boldsymbol{\varepsilon}^\star} = \mathrm{DFT}(\boldsymbol{\varepsilon}^\star), \tag{4.24}$$

$$\hat{\boldsymbol{\varepsilon}^\star} = -\hat{\mathbb{G}}_0[\hat{\boldsymbol{p}}], \qquad \hat{\boldsymbol{p}} = -\hat{\mathbb{L}}_0[\hat{\boldsymbol{\varepsilon}^\star}], \tag{4.25}$$

$$\hat{\boldsymbol{\varepsilon}^\star}(0) = \bar{\boldsymbol{\varepsilon}}, \qquad \hat{\boldsymbol{p}}(0) = \bar{\boldsymbol{\sigma}}, \tag{4.26}$$

$$\boldsymbol{\varepsilon}^{n+1} = \mathrm{DFT}^{-1}(\hat{\boldsymbol{\varepsilon}^\star}), \qquad \boldsymbol{\sigma}^{n+1} = \mathrm{DFT}^{-1}(\hat{\boldsymbol{p}}). \tag{4.27}$$

In these equations, the sign of the eigenstrain $\boldsymbol{\varepsilon}^\star$ is changed, to get an identical algorithm for the strain and stress formulation. Prescriptions for $\hat{\mathbb{G}}_0$ can be found in Mura (1987) for different types of anisotropy.

For the computation of the Hashin-Shtrikman bounds (4.16) and (4.19), the Green operators $\mathbb{G}_0$ and $\mathbb{L}_0$ in the basic scheme are replaced by their periodized counterparts $\mathbb{G}_0^\#$ and $\mathbb{L}_0^\#$ (Brisard and Dormieux, 2010).

Hashin-Shtrikman bounds on the effective moduli

According to Kabel and Andrä (2013), it is possible to use the solutions of the Lippmann-Schwinger equations (4.16) and (4.19) to obtain upper and lower bounds on the effective moduli. Due to the assumptions of the Hashin-Shtrikman bounds on $\mathbb{C}_0$, it is not possible to calculate an upper (lower) bound of the effective stiffness numerically, if the composite contains any rigid (porous) region. Since the upper Hashin-Shtrikman bound depicts a rigid behavior in the first case, and the lower Hashin-Shtrikman bound a material without any elastic stiffness in the second, this is not a real limitation. If a composite contains rigid and also porous regions, the assumptions of the Hashin-Shtrikman bounds cannot be fulfilled.

In can be shown (Müller et al., 2015b), that combining (4.10) and (4.12) with (4.14) yields

$$\bar{\boldsymbol{\varepsilon}} \cdot \bar{\mathbb{C}}[\bar{\boldsymbol{\varepsilon}}] = \bar{\boldsymbol{\varepsilon}} \cdot \mathbb{C}^{\mathsf{HS}}_{\boldsymbol{\varepsilon}}[\bar{\boldsymbol{\varepsilon}}]. \tag{4.28}$$

It follows

$$\mathbb{C} \underset{(\geq)}{\leq} \mathbb{C}_0 \Rightarrow \bar{\mathbb{C}} \underset{(\geq)}{\leq} \mathbb{C}^{\mathsf{HS}}_{\boldsymbol{\varepsilon}}. \tag{4.29}$$

Further it can be shown, that combining (4.11) and (4.13) with (4.18) yields an analogous result for the compliance

$$\bar{\boldsymbol{\sigma}} \cdot \bar{\mathbb{S}}[\bar{\boldsymbol{\sigma}}] - \bar{\boldsymbol{\sigma}} \cdot \mathbb{S}_0[\bar{\boldsymbol{\sigma}}] = \bar{\boldsymbol{\sigma}} \cdot \left(\mathbb{S}_0 - \mathbb{S}^{\mathsf{HS}}_{\boldsymbol{\sigma}}\right)[\bar{\boldsymbol{\sigma}}]. \tag{4.30}$$

Therefore, if $\mathbb{S} \underset{(\leq)}{\geq} \mathbb{S}_0$

$$\bar{\mathbb{S}} \underset{(\geq)}{\leq} 2\mathbb{S}_0 - \mathbb{S}^{\mathsf{HS}}_{\boldsymbol{\sigma}} \underset{(\geq)}{\leq} 2\bar{\mathbb{S}} - \mathbb{S}^{\mathsf{HS}}_{\boldsymbol{\sigma}}, \tag{4.31}$$

which means that

$$\mathbb{S} \underset{(\leq)}{\geq} \mathbb{S}_0 \Rightarrow \mathbb{S}^{\mathsf{HS}}_{\boldsymbol{\sigma}} \underset{(\geq)}{\leq} \bar{\mathbb{S}}. \tag{4.32}$$

For non-rigid materials this implies

$$\mathbb{C} \underset{(\geq)}{\leq} \mathbb{C}_0 \Rightarrow \bar{\mathbb{C}} \underset{(\geq)}{\leq} \mathbb{C}^{\mathsf{HS}}_{\boldsymbol{\sigma}}. \tag{4.33}$$

According to Michel et al. (2001) and Eyre and Milton (1999), the convergence of the Neumann series expansion in equation (4.20) for the strain based formulation of the Lippmann-Schwinger equation (4.16) is only guaranteed for $\mathbb{C} \leq \mathbb{C}_0$. For the stress based formulation only $\mathbb{S} \leq \mathbb{S}_0$ guarantees the convergence of the Neumann expansion (4.20).

Thus, using the strain (stress) based formulation it is possible to determine an upper (lower) bound of the effective stiffness if the composite does not contain any rigid region (porous region)

$$\mathbb{C}^{\mathsf{HS}}_{\boldsymbol{\sigma}} \leq \bar{\mathbb{C}} \leq \mathbb{C}^{\mathsf{HS}}_{\boldsymbol{\varepsilon}}. \tag{4.34}$$

4.2.3 Mean-field homogenization

The FFT based full-field methods described in the foregoing section are compared with the SC, the IDD and the TS approach, which are presented in the sections 3.4.2, 3.4.3 and 3.4.4.

4.3 Model microstructures and material parameters

4.3.1 Generation of the model microstructures

The mean orientation distribution can be described with fabric tensors of second rank $\boldsymbol{N}$ (Kanatani, 1984), also called fiber orientation tensors (Advani and Tucker III, 1987). For N equal weighted fiber orientations $\boldsymbol{n}$, the orientation tensor is defined as

$$\boldsymbol{N} = \frac{1}{N} \sum_{\alpha=1}^{N} \boldsymbol{n}_\alpha \otimes \boldsymbol{n}_\alpha. \tag{4.35}$$

Since the fiber axis orientation $\boldsymbol{n}$ is normalized, the trace of the fiber orientation tensor is always one: $\mathrm{sp}(\boldsymbol{N}) = 1$.

Three different microstructures have been used for the comparison of the methods. Firstly, a microstructure with unidirectional aligned fibers has been considered. Results affiliated with this microstructure are referred to with UD in the following. Secondly, inspired by the orientation distribution of the fiber axes in injection-molded thin plates, two different microstructures with misaligned fiber axes have been considered. Results based on this microstructures are denoted with TP1 and TP2 (thin plate) in the following. All microstructures are generated under a periodicity constraint. In Table 4.1, the components of the orientation tensors, which have been used to generate the model microstructures, are given. The artificial microstructures have been generated by using Math2Market (2014). The corresponding data files can be found on the internet page `http://www.itm.kit.edu/cm/288.php`. The algorithm is described in Schladitz et al. (2006).

Table 4.1: Components of orientation tensors, used for the generation of the model microstructures

	N_{11}	N_{22}	N_{33}	N_{23}	N_{13}	N_{12}
$\boldsymbol{N}_0^{\mathrm{TP1}}$	0.61	0.36	0.03	0.0	0.0	0.0
$\boldsymbol{N}_0^{\mathrm{TP2}}$	0.80	0.18	0.02	0.0	0.0	0.0
$\boldsymbol{N}_0^{\mathrm{UD}}$	1.0	0.0	0.0	0.0	0.0	0.0

For all microstructures, the fibers have been modeled by cylinders with a length of $l = 200\,\mu\mathrm{m}$ and a diameter of $d = 10\,\mu\mathrm{m}$. Ten different RVEs have been realized for each microstructure type. The side length of the RVEs has been $d_{\mathsf{RVE}} = 250\,\mu\mathrm{m}$, and they have been discretized with the resolutions 125, 250 and 500, corresponding to a side length of one voxel of $2\,\mu\mathrm{m}$, $1\,\mu\mathrm{m}$, and $0.5\,\mu\mathrm{m}$, in all directions for consideration in the full-field approach. For all realizations, the fiber volume fraction is assumed to be equal to $c_{\mathsf{F}} = 13\%$. Additionally, for UD, fiber volume fractions of 17% and 21% have been taken into account. Since the mean-field methods only rely on the orientation information, the explicit geometrical description of the orientation and geometry of each fiber is necessary. In Figure 4.1, one example for UD each of the microstructures is given.

4.3.2 Properties of the model microstructures

The microstructure generation process, which is implemented in GeoDict, is a random process.[3] In the UD case, the microstructure generation algorithm matches perfectly the unidirectional alignment of the fibers. The achieved mean orientation tensors of the TP

[3] The model microstructures have been generated by M. Kabel (Fraunhofer ITWM).

(a)

(b)

Figure 4.1: (a) UD microstructure; (b) TP microstructure.

RVEs are compared with the target orientation tensors in Table 4.2.

Table 4.2: Components of the average orientation tensors of 10 realization of the TP microstructures and their deviations compared with the target orientation tensors (see Table 4.1)

	$\bar{\boldsymbol{N}}^{\mathsf{TP1}}$	$\boldsymbol{N}_0^{\mathsf{TP1}} - \bar{\boldsymbol{N}}^{\mathsf{TP1}}$	$\bar{\boldsymbol{N}}^{\mathsf{TP2}}$	$\boldsymbol{N}_0^{\mathsf{TP2}} - \bar{\boldsymbol{N}}^{\mathsf{TP2}}$
N_{11}	0.6083	0.17%	0.8058	-0.58%
N_{22}	0.3615	-0.15%	0.1743	0.57%
N_{33}	0.0302	1.02%	0.0199	0.01%
N_{23}	0.0064	-0.64%	-0.0006	0.06%
N_{13}	-0.0017	0.17%	0.0022	-0.22%
N_{12}	0.0004	-0.04%	0.0114	-0.11%

4.3.3 Parameter overview

As mentioned above, three different microstructure types with three different fiber volume fractions have been regarded for the comparison of the homogenization methods. Additionally, three different combination of elastic moduli for fibers and matrix have been taken into account: First, the elastic moduli have been chosen corresponding to polypropylene ($E_{\mathsf{PP}} = 1.665\,\mathrm{GPa}$, $\nu_{\mathsf{PP}} = 0.36$) and glass ($E_{\mathsf{G}} = 73\,\mathrm{GPa}$, $\nu_{\mathsf{G}} = 0.2$) (Joshi et al., 1994; Tomasetti et al., 1998). These materials are frequently used to compound composites like PPGF30, consisting of polypropylene reinforced with glass fiber of a weight fraction of 30%. The phase contrast of this combination amounts to $\xi = 44$.

Further, the phase contrast has been increased to 100 and 1000, whereby Young's modulus of the matrix and fibers is set to

$E_{\mathsf{M}} = 1\,\mathrm{GPa}$ and $E_{\mathsf{F}} = 100\,\mathrm{GPa}$ or $E_{\mathsf{F}} = 1000\,\mathrm{GPa}$. The parameter variation is summarized in Table 4.3.

Table 4.3: Variation of microstructural and material parameters

Microstr.	Fiber vol. fraction		Material combination			
TP1	TP1/TP2	UD	ξ	44	100	1000
TP2	13%	13%	E_{M} [GPa]	1.665	1	1
UD		17%	ν_{M}	0.36	0.36	0.36
		21%	E_{F} [GPa]	73	100	1000
			ν_{F}	0.2	0.2	0.2

4.4 Numerical results[4]

4.4.1 Directional dependent elastic properties

To compare the effective stiffnesses, the directional dependent Young's modulus and bulk modulus can be evaluated. The directional dependent Young's modulus

$$\frac{1}{E(\boldsymbol{d})} = \boldsymbol{d} \otimes \boldsymbol{d} \cdot \bar{\mathbb{S}}\,[\boldsymbol{d} \otimes \boldsymbol{d}] \tag{4.36}$$

and the directional dependent bulk modulus

$$\frac{1}{3K(\boldsymbol{d})} = \boldsymbol{I} \cdot \bar{\mathbb{S}}\,[\boldsymbol{d} \otimes \boldsymbol{d}] \tag{4.37}$$

determine uniquely the effective compliance tensor $\bar{\mathbb{S}} = \bar{\mathbb{C}}^{-1}$ (see, e.g., Böhlke and Brüggemann (2001)). In Figure 4.2, the graphical

[4] The numerical homogenization results have been provided by M. Kabel (Fraunhofer ITWM).

representations of these quantities are given exemplarily for a UD realization. In the following, the numerical results are compared by

(a) (b)

Figure 4.2: Graphical representation of the directional dependent (a) Young's modulus and (b) bulk modulus for one UD realization.

means of the directional dependent Young's modulus. on the one hand, a quarter of the contour of this quantity is shown on the x-y-plane. Here, this plane is the plane with the main fiber orientation distribution in terms of UD and TP. On the over hand, the comparison is expressed by means of a relative deviation of the directional dependent Young's modulus of method X and Y:

$$\eta_{\mathsf{Y}}^{\mathsf{X}} = \left(\frac{\int_S \left(E^{\mathsf{X}}(\boldsymbol{n}) - E^{\mathsf{Y}}(\boldsymbol{n}) \right)^2 \, \mathrm{d}S}{\int_S \left(E^{\mathsf{X}}(\boldsymbol{n}) \right)^2 \, \mathrm{d}S} \right)^{1/2} . \tag{4.38}$$

4.4.2 Resolution, size and realization dependency of Young's modulus

Spatial resolution of RVE In order to evaluate the dependencies of the FFT solution on the resolution, Young's moduli have been calculated in $0°$-direction (E_0) for each realization of the three different resolutions for two different phase contrasts, $\xi = 44$ and $\xi = 1000$. For UD microstructures with a volume fraction of 13%, these results are shown in Figure 4.3, and, for TP1 microstructures in Figure 4.4. In both figures, the mean Young's modulus values and the corresponding standard deviation of all realizations for the lower bound (FFT-HS−), the basic scheme (FFT-BS), and the upper bound (FFT-HS+) in the appropriate direction are shown.

Since the averaged Young's modulus values of the FFT-BS solution are on an approximately constant level for both microstructure types, a dependency on the resolution cannot be observed. The standard deviation of the FFT-BS results in terms of different RVE realizations is similar for all resolutions. Contrary to this, FFT-HS+ and FFT-HS− dependent obviously on the resolution: the gap between the corresponding bounds is narrowed with increasing resolution. This can be reasoned with an increasing space for solution in this case of voxel-wise constant field variables. Thus, tighter bounds can be found based on equation (4.18).

For the smaller phase contrast of $\xi = 44$, FFT-HS+ is closer and converges faster to FFT-BS for UD and TP1 in the considered direction. Just the opposite can be observed for the larger phase contrast of $\xi = 1000$. The standard deviation of FFT-HS+ is slightly increasing for UD with $\xi = 44$ and obviously increasing for UD with $\xi = 1000$. In the TP1 case, its standard deviation is approximately constant for the smaller phase contrast and decreasing for the larger phase contrast. The standard deviation of FFT-HS− increases in all cases with increasing resolution.

Size of RVE In order to verify the representativity of the volume element for the misaligned microstructure configuration, additional volume elements of three different sizes have been considered for TP1: Starting with $250\,\mu m$, the side length has been doubled two

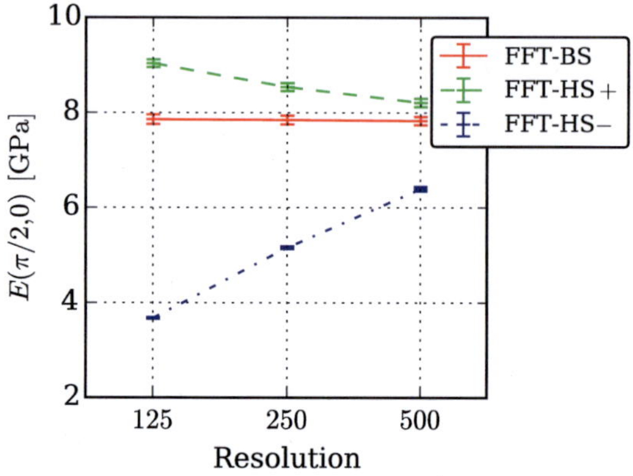

(a) UD with $c_F = 13\%$ and $\xi = 44$

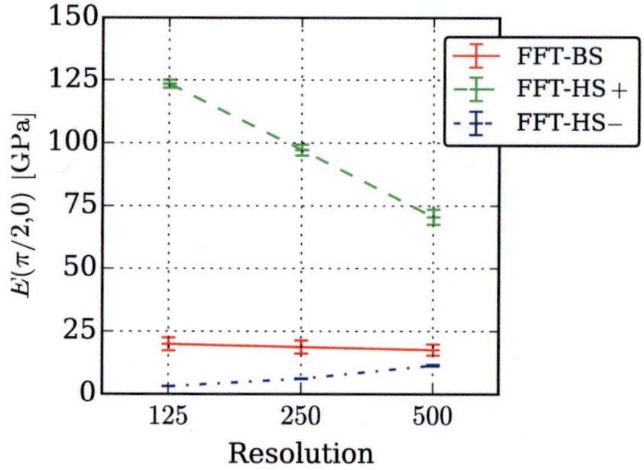

(b) UD with $c_F = 13\%$ and $\xi = 1000$

Figure 4.3: Mean Young's modulus and standard deviation in 0°-direction of UD microstructure realizations in dependence of the resolution of the RVE for FFT methods

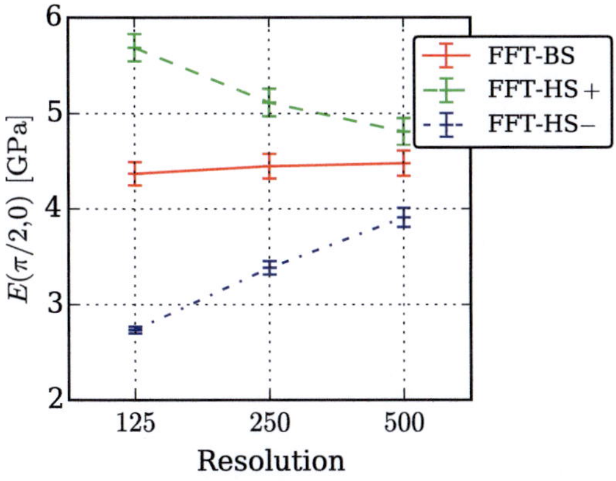

(a) TP1 with $c_F = 13\%$ and $\xi = 44$

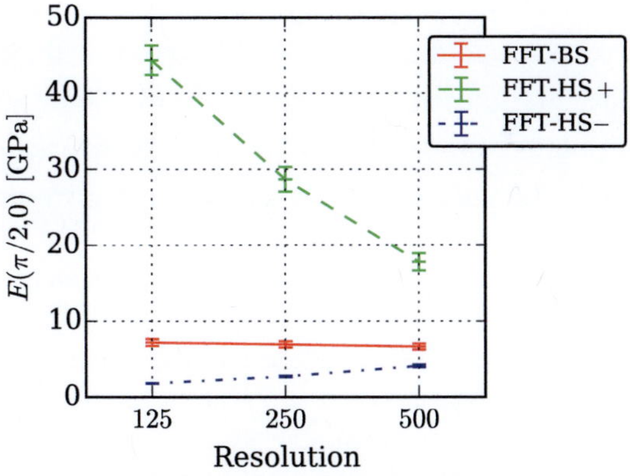

(b) TP1 with $c_F = 13\%$ and $\xi = 1000$

Figure 4.4: Mean Young's modulus and standard deviation in $0°$-direction of TP1 microstructure realizations in dependence of the resolution of the RVE for FFT methods times to $500\,\mu m$ and $1000\,\mu m$. Again, ten volume elements with a resolution of 125, 250 and 500 voxels, respectively, in each direction have been generated and homogenized using the FFT and MF methods. The resulting averaged Young's modulus in $0°$-direction and its standard deviation are shown in Figure 4.5(a) for the FFT and in Figure 4.5(b) for the MF approaches.

Despite the low resolution which has been chosen due to computational issues, it is obvious from Figure 4.5, that the mean values of Young's modulus are approximately constant for all RVE sizes. The FFT and MF methods show qualitatively comparable results in Figure 4.5(a) and in Figure 4.5(b), respectively. On the one hand, for small RVEs with 130 inclusions, a dependency on the specific realization is evident. On the other hand, for the midsize and large volume elements with 1035 and 8277 inclusions, respectively, Young's modulus does not show a significant dependency on a specific realization. The correspondence of the orientation distribution in terms of the fiber orientation tensor compared to the target given in Table 4.1 is the better, the larger the volume elements become, see Table 4.4. Since the mean stiffnesses and also the mean direction-

Table 4.4: Components of orientation tensors of RVEs with an endge-length of $250\,\mu m$, $500\,\mu m$ and $1000\,\mu m$

	N_{11}	N_{22}	N_{33}	N_{23}	N_{13}	N_{12}
$\boldsymbol{N}^{TP1}_{130}$	0.6086	0.3614	0.0299	0.0062	−0.0020	0.0014
$\boldsymbol{N}^{TP1}_{1035}$	0.6115	0.3581	0.0304	−0.0005	0.0007	−0.0007
$\boldsymbol{N}^{TP1}_{8277}$	0.6113	0.3586	0.0301	0.0000	−0.0001	−0.0003

dependent Young's modulus are constant regarding different RVE sizes, it is concluded, that the representativity is also given for the small RVEs in an averaged manner.

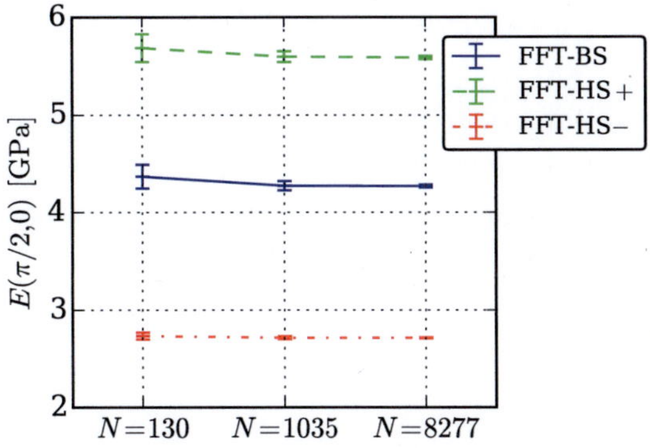

(a) TP1 with $c_F = 13\%$ and $\xi = 44$

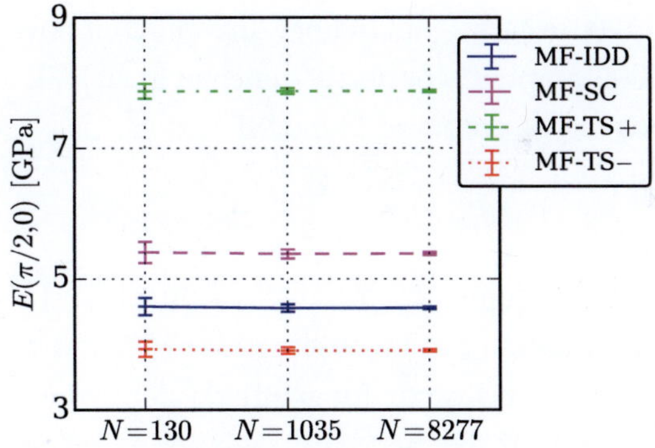

(b) TP1 with $c_F = 13\%$ and $\xi = 44$

Figure 4.5: Mean Young's modulus and standard deviation of TP1 microstructure realizations in $0°$-direction in dependence of the size of the RVE for (a) FFT methods and (b) MF approaches

Variation in RVE generation The variation of the microstructure realization process is analyzed by means of averaged directional dependent Young's moduli and standard deviations of all realizations on the x-y-plane. In Figure 4.6 and 4.7, FFT results are shown for UD and TP1, respectively. In each figure, results for microstructures with a fiber volume fraction of 13% and a phase contrast of 44 and 1000 are presented. Especially for UD, a direction dependence of the standard deviation of Young's modulus can be observed: The greatest deviation occurs in fiber direction ($0°$-direction) and the smallest in transversal direction for all FFT methods. It increases for higher averaged Young's modulus. Thus, for the upper bound FFT-HS+ and for larger phase contrast the greatest standard deviation is found. However, relating the standard deviation to the averaged Young's modulus in the corresponding direction reveals two facts: The greatest relative deviation of approximately 2% for $\xi = 44$ and 25% for $\xi = 1000$ occurs again in fiber direction but now for FFT-BS. Since the UD alignment is perfectly matched in all RVE realizations, no deviations can be observed in the MF results.

In the TP1 case in Figure 4.7, no distinct direction dependence of the standard deviation can be observed for FFT results. However, the greatest deviations occur for methods delivering the greatest Young's modulus. If the standard deviation is related to the averaged Young's modulus, a deviation of 8% for $\xi = 44$ and 18% for $\xi = 1000$ is found. The MF results also depend on the RVE realization in the TP1 case, see Figure 4.8. Here, again, no distinct direction dependence of the standard deviation can be observed. MF-SC delivers the highest sensitivity on specific RVE realizations and, thus, the highest standard deviation, also compared to FFT methods: 11% for $\xi = 44$ and 25% for $\xi = 1000$.

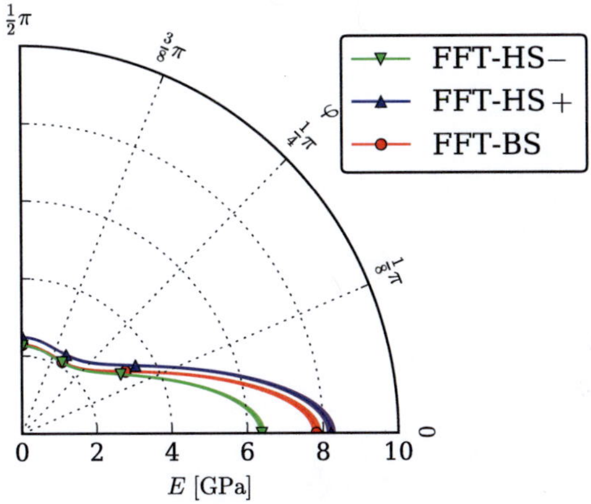

(a) UD with $c_F = 13\%$ and $\xi = 44$

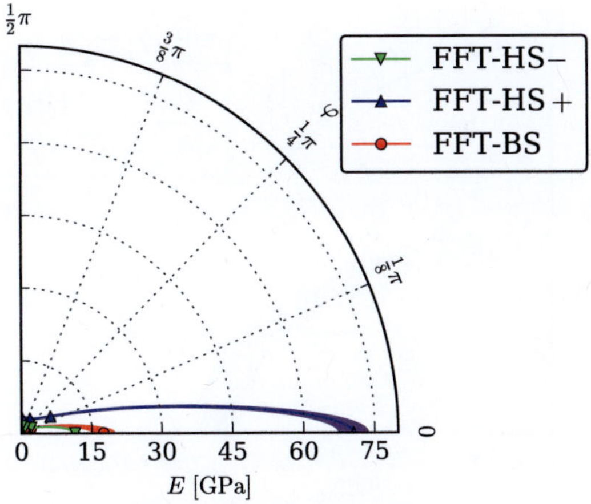

(b) UD with $c_F = 13\%$ and $\xi = 1000$

Figure 4.6: Direction dependent Young's modulus and standard deviation in x-y-plane for UD microstructure realizations for FFT methods

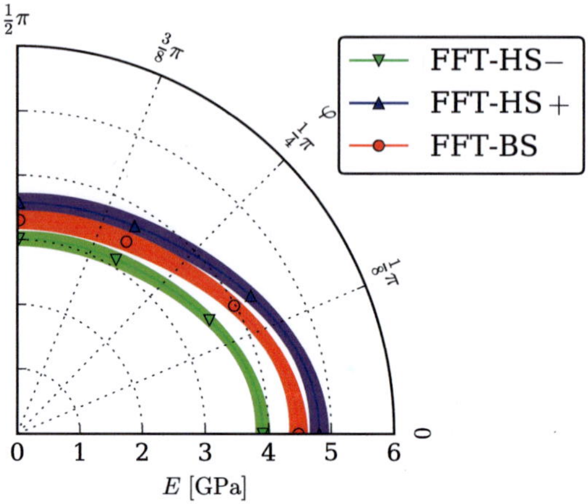

(a) TP1 with $c_F = 13\%$ and $\xi = 44$

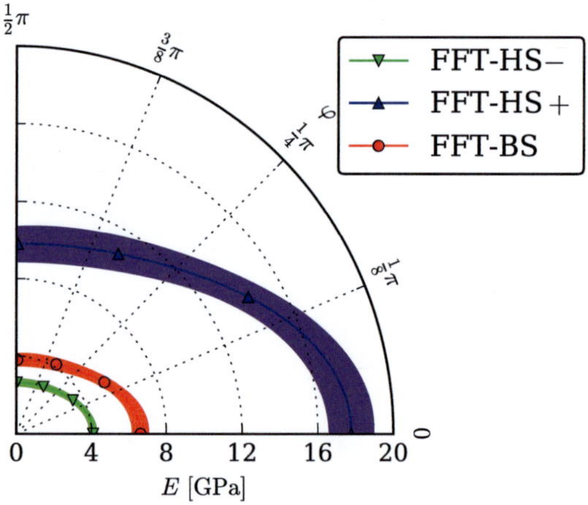

(b) TP1 with $c_F = 13\%$ and $\xi = 1000$

Figure 4.7: Direction dependent Young's modulus and standard deviation in x-y-plane for TP1 microstructure realizations for FFT methods

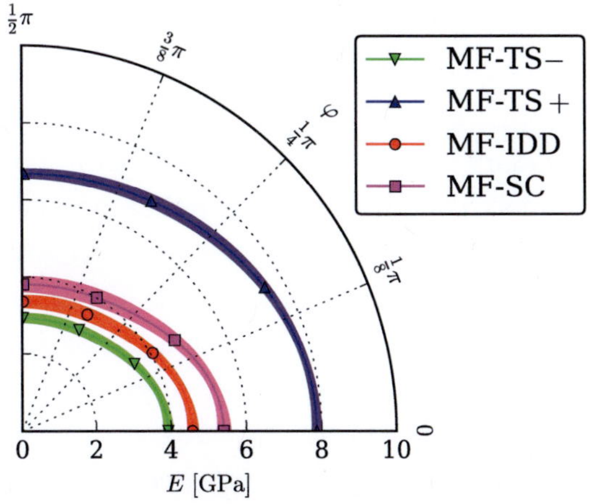

(a) TP1 with $c_F = 13\%$ and $\xi = 44$

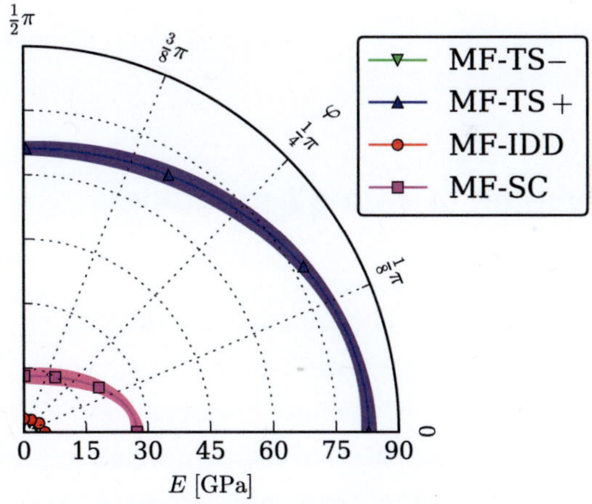

(b) TP1 with $c_F = 13\%$ and $\xi = 1000$

Figure 4.8: Direction dependent Young's modulus and standard deviation in x-y-plane for TP1 microstructure realizations for MF methods

Effects of the periodicity constraint For the generation of the TP microstructures, a periodicity constraint has been imposed. In order to quantify the effect of the periodicity, non-periodic TP1 RVEs with the side length of 250, 500 and $1000\,\mu\mathrm{m}$ have been generated additionally. Although the RVEs are not periodic, it is still possible to solve the boundary value problem using the FFT approaches with periodic boundary conditions. Compared to the results of FFT-BS for periodic microstructures, the relative deviation, which is defined in equation (4.38), decreases with increasing RVE side length of the non-periodic microstructures. For $250\,\mu\mathrm{m}$ the deviation amounts to 9.0%, for $500\,\mu\mathrm{m}$ to 6.7% and for the largest RVE with $1000\,\mu\mathrm{m}$ to 4.8%. The periodicity constraint ensures a constant fiber length, however, this is not the case for the non-periodic RVEs, which contain shorter fiber ends at their boundaries. Since the volume fraction of the short fiber ends decreases for increasing RVE size, the results are reasonable.

4.4.3 Deviation of Young's modulus

In the following, FFT-BS is compared with MF-IDD and MF-SC by means of a relative deviation defined in equation (4.38). Herein, the highest resolution of the RVEs have been used for FFT-BS. The effective averaged Young's moduli of the mentioned methods are compared for all microstructure types, phase contrasts and fiber volume fractions. For each comparison, the relative deviation has been calculated and shown in a heat map, see, e.g., Figure 4.9. Herein, the relative deviation is entered directly on the one hand, and represented in color in order to see the tendencies better on the other hand.

In Figure 4.9(a), first, FFT-BS and MF-IDD are compared for all microstructure types and phase contrasts. The smallest deviation of 1.5% occurs for TP1 with the smallest phase contrast $\xi = 44$. The largest deviation of 31.8% is found for UD with $\xi = 1000$. The deviation increases with increasing phase contrast and with advancing alignment of the fibers (TP1 $\rightarrow$ TP2 $\rightarrow$ UD). Similar results can be seen in Figure 4.9(c), where FFT-BS and MF-IDD are compared for UD microstructures with all considered phase contrasts and volume fractions. Starting from 3%, the deviation increases with increasing phase contrast and also with increasing volume fraction up to 49.5%.

The relative deviation between MF-SC and FFT-BS for UD, shown in Figure 4.9(d), reveal the same tendencies as MF-IDD and FFT-BS. Again, the deviation increases with increasing fiber volume fraction and phase contrast. The minimum and maximum deviations amount to 4.5% and 29.2%, respectively. Another tendency can be observed for the comparison of MF-SC and FFT-BS in terms of different microstructure types, see Figur 4.9(b). Here, the deviation still increases with increasing phase contrast, however, it decreases with advancing alignment of the fibers.

The direct comparison of Young's modulus for TP2 with $c_{\mathsf{F}} = 13\%$ and $\xi = 100$ is shown in Figure 4.10(a) on the x-y-plane for all MF and FFT methods. Accordingly, in Figure 4.10(b), the UD results of all methods are compared for $c_{\mathsf{F}} = 17\%$ and $\xi = 100$. It can be seen, that the results of all methods are located between MF-TS+ and FFT-HS−, where FFT-BS appears between the narrow FFT bounds. MF-SC predicts a higher Young's modulus than FFT-BS and a partially higher Young's modulus than FFT-HS+, especially in fiber direction in the UD case. MF-IDD is located between FFT-BS and the lower bound FFT-HS−. In the UD case, MF-IDD and MF-TS− coin-

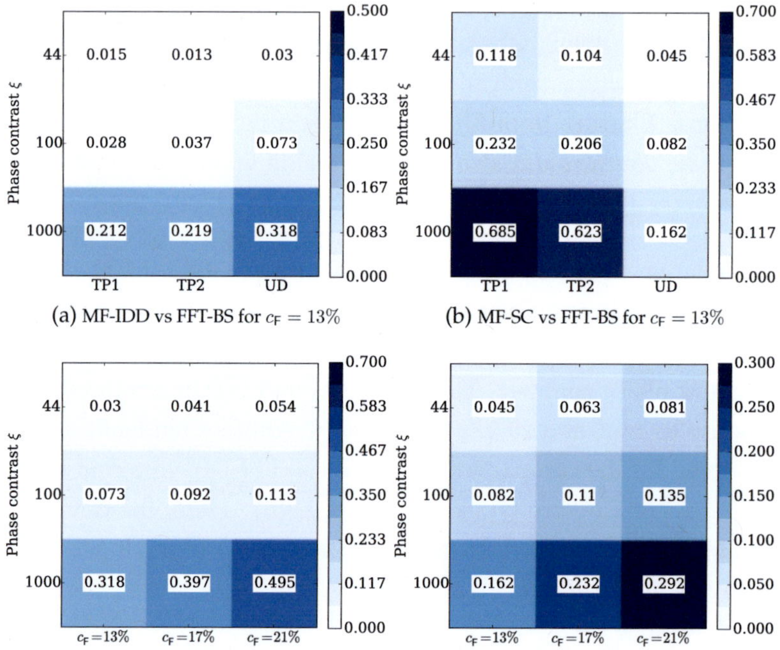

(a) MF-IDD vs FFT-BS for $c_F = 13\%$

(b) MF-SC vs FFT-BS for $c_F = 13\%$

(c) MF-IDD vs FFT-BS for UD

(d) MF-SC vs FFT-BS for UD

Figure 4.9: Relative deviation of FFT-BS and (a) MF-IDD for TP1, TP2 and UD with $c_F = 13\%$, (b) MF-SC for TP1, TP2 and UD with $c_F = 13\%$, (c) MF-IDD for UD with $c_F = 13, 17, 21\%$ and (d) MF-SC for UD with $c_F = 13, 17, 21\%$

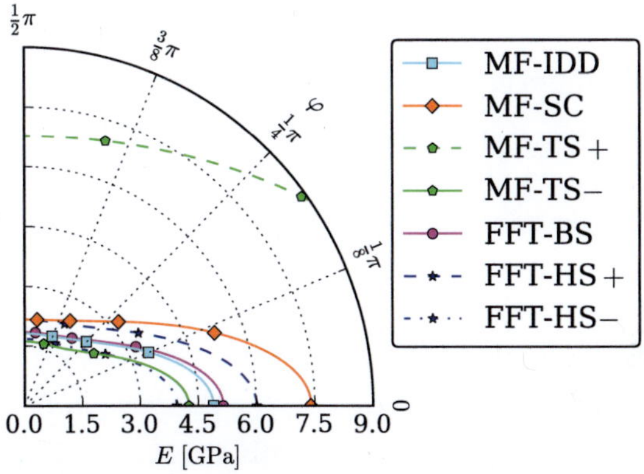

(a) TP2 with $c_F = 13\%$ and $\xi = 100$

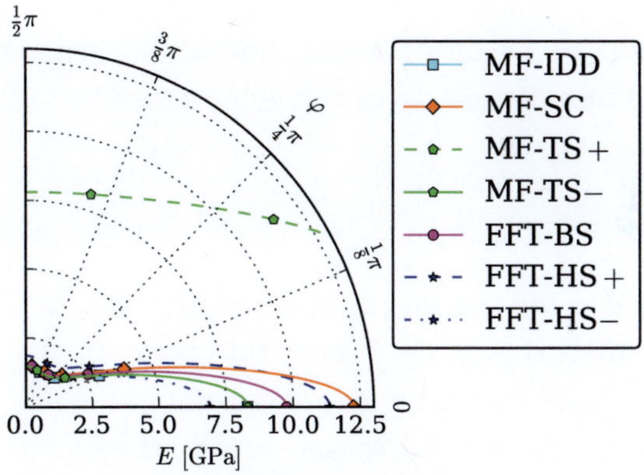

(b) UD with $c_F = 17\%$ and $\xi = 100$

Figure 4.10: Direction dependent Young's modulus of all FFT and MF methods on x-y-plane

Figure 4.11: Approximation of a cylindrical fiber with equal volume and aspect ratio or volume and length

cide for all parameters and appear between FFT-BS and FFT-HS−. MF-TS+ predicts a much higher Young's modulus then all other methods, additionally for UD, the shape of the MF-TS+ Young's modulus is different compared to the other approaches.

Effects of the ellipsoidal approximation of the cylindrical fibers

Generally, it is known that for UD, the effective material behavior in fiber direction is getting stiffer with increasing fiber length. The MF results discussed so far are based on the modeling of the cylindrical fibers by ellipsoids of equal aspect ratio and volume. Thus, the ellipsoids are longer than the cylindrical fibers in the RVE. The half axes of the ellipsoid, a_1, a_2 and a_3 are scaled by $(3/2)^{1/3}$:

$$2a_1 = \left(\frac{3}{2}\right)^{1/3} l, \qquad 2a_2 = 2a_3 = \left(\frac{3}{2}\right)^{1/3} d,$$

where l and d are the length and the diameter of the cylindrical fiber. Alternatively, the cylindrical fibers can also be approximated by ellipsoids with the side condition of equal length and equal volume. Herein, the second and third half axis have to be scaled with $(3/2)^{1/2}$, while the first is exactly $2a_1 = l$. In Figure 4.11, both approximations are shown for a fiber with an aspect ratio of 20. The ellipsoidal approximation with equal volume and aspect ratio overestimates the length and the width of the cylinder by about 14.5%, while the approximation with equal volume and equal length overestimates the width by 22.5%. In the following, these two approaches to model the cylindrical fibers by ellipsoids

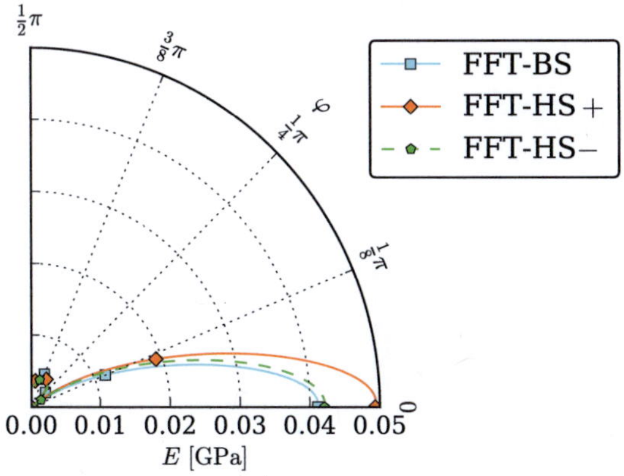

Figure 4.12: Relative deviation of the directional dependent Young's modulus $\left|E^{X}_{cyl}(\varphi) - E^{X}_{ell}(\varphi)\right| / E^{X}_{cyl}(\varphi)$ for UD with $c_F = 13\%$ and $\xi = 44$ of the FFT methods for ellipsoidal and cylindrical inclusion geometry

are distinguished by the terms first type ellipsoid and second type ellipsoid.

To determine the effect of the ellipsoidal approximation on the effective material properties, additional RVEs with unidirectional aligned first type ellipsoids and a fiber volume fraction of $c_F = 13\%$ have been generated. In Figure 4.12, the relative differences between the averaged stiffnesses of the RVEs with cylindrical inclusions and ellipsoidal inclusions for FFT-BS, FFT-HS+ and FFT-HS− are shown for the phase contrast of 44. In this figure, that error is indicated, which is introduced if the cylindrical fibers are modeled by first type ellipsoids. In the fiber direction, the difference ranges between 4% in case of FFT-BS and FFT-HS−, and 5% in case of FFT-HS+.

Using the second type ellipsoid for MF-SC and MF-IDD approaches, denoted by MF-SC2 and MF-IDD2, again an analysis of

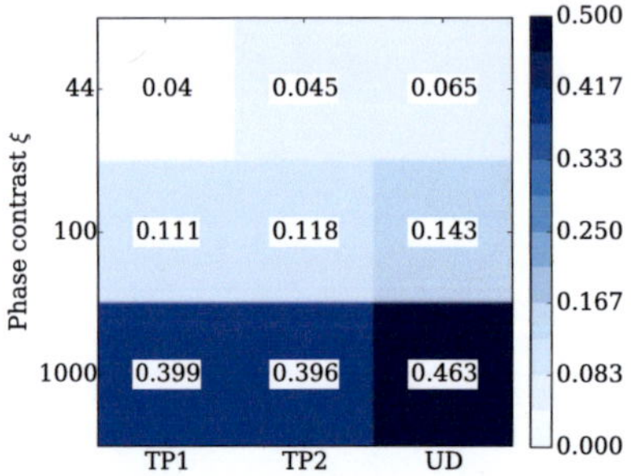

(a) MF-IDD2 vs FFT-BS for $c_F = 13\%$

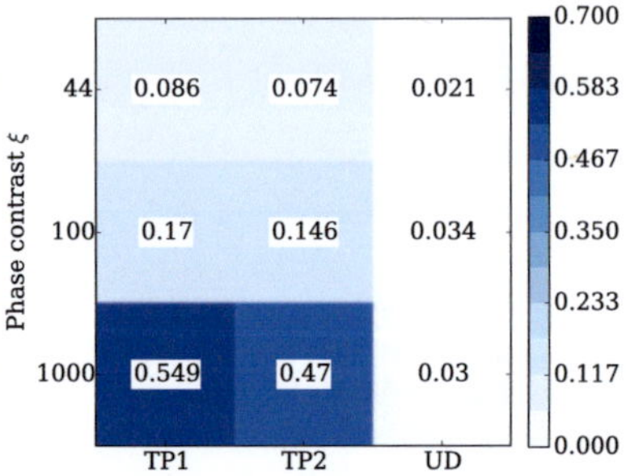

(b) MF-SC2 vs FFT-BS for $c_F = 13\%$

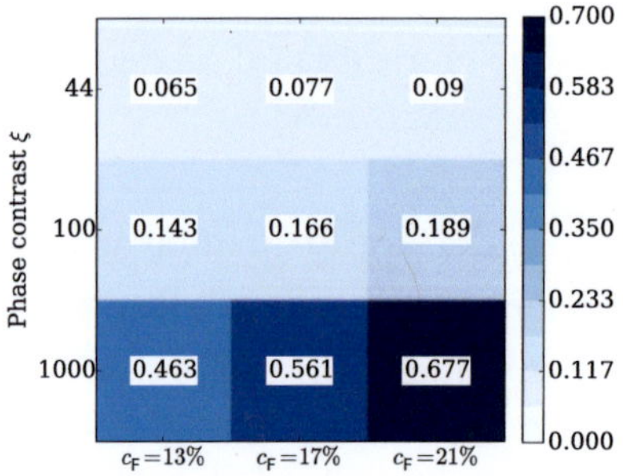

(c) MF-IDD2 vs FFT-BS for UD

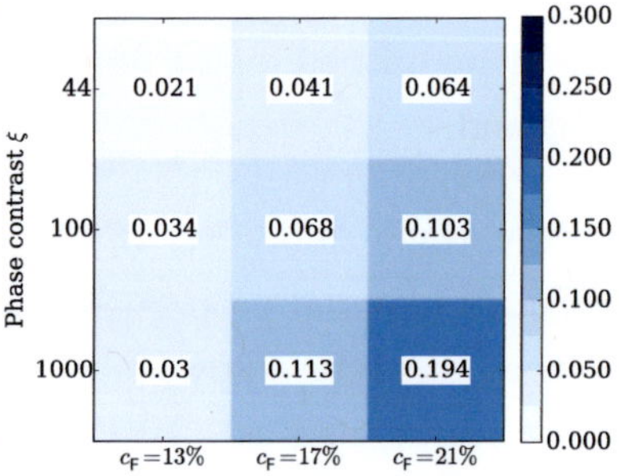

(d) MF-SC2 vs FFT-BS for UD

Figure 4.13: Relative deviation of FFT-BS and (a) MF-IDD2 for TP1, TP2 and UD with $c_F = 13\%$, (b) MF-SC2 for TP1, TP2 and UD with $c_F = 13\%$, (c) MF-IDD2 for UD with $c_F = 13, 17, 21\%$ and (d) MF-SC2 for UD with $c_F = 13, 17, 21\%$

the relative deviation to FFT-BS with cylindrical inclusions have been performed. The results of this analysis is shown in Figure 4.13. Compared with the former deviation analysis shown in Figure 4.9, it can be seen, that the tendencies remain the same. However, all deviations of the comparison of MF-IDD2 and FFT-BS are higher than the deviations between MF-IDD and FFT-BS. The contrary is the case for MF-SC2 and FFT-BS: Here, the deviation are smaller than in the former analysis. This can be reasoned by the smaller aspect-ratio of the second type ellipsoidal approximation. Generally, MF-IDD predicts smaller Young's moduli than FFT-BS. Smaller aspect-ratios lead to smaller Young's moduli and, therefore, the deviations become larger. In case of MF-SC, which generally predicts larger Young's moduli than FFT-BS, the deviations become smaller for smaller aspect-ratios of the inclusions.

4.5 Summary and conclusions

The main objective of this chapter is the detailed comparison of the predicted effective linear elastic properties of full-field and mean-field methods for discontinuous short-fiber reinforced composites. For this purpose, the IDD scheme, the SC method and a TS approach, on the one hand, and a full-field solution based on fast Fourier transformation, on the other hand, have been applied. In the case of the mean-field methods, an explicit geometrical description of the fibers in terms of their geometrical properties can be taken into account. Whereas for the full-field method, a regular three-dimensional discretization of the microstructures with voxel-wise constant properties is additionally necessary. The explicit description of the microstructures is available online. The properties of the considered microstructures have been analyzed. Three different types of microstructures have been considered, two misaligned and one unidirectional configuration, in conjunction with three phase contrasts, the smallest contrast of 44 for a PPGF composite and two artificial of 100 and 1000. Three different fiber volume fractions have been taken into account for the unidirectional configuration: 13%, 17% and 21%, what corresponds to weight fractions of 30%, 40% and 50% for the PPGF composite material. The numerical solutions have been compared in terms of the directional dependent Young's modulus regarding the dependency on resolution and size of the RVEs, the relative deviations of Young's modulus, and the anisotropy of Young's modulus in the plane of the main fiber orientation distribution.

In the following, the term *Young's modulus* always refers to the realization averaged Young's modulus. The term *specific Young's modulus* denotes Young's modulus of one particular RVE realization.

With respect to the model microstructures and their discretization, the following can be concluded:

- The range of Young's modulus between the upper and lower HS bounds of the FFT solution depends on the resolution (voxel) of the RVE. With increasing resolution, both bounds converge to the FFT-BS solution. The Young's modulus of the FFT-BS solution does not show an obvious dependency on the selected resolution of the RVE.

- No significant effect of the size of the RVE on Young's modulus of the TP configuration has been observed. While for small RVEs with 130 fibers, the specific Young's modulus depends on the realization, this is not the case for larger RVEs with 1035 and more fibers.

- In the UD case, the standard deviation of specific Young's moduli of FFT solutions show a dependency on a particular realization only in fiber direction. The deviation is higher for higher effective Young's moduli. The highest relative deviation, where the standard deviation is divided by the average Young's modulus, occurs for FFT-BS: 2% for a phase contrast of 44 and 25% for a phase contrast of 1000. No deviations occur for MF approaches.

- In the TP1 case, no distinct direction dependency of the standard deviation of specific Young's moduli can be observed. The SC method shows the highest sensitivity on the specific RVE realization. The relative standard deviation amounts to 11 and 25% for a phase contrast of 44 and 1000, respectively.

Thus, an RVE with a side length of $250\,\mu\text{m}$ is sufficient for the considered microstructure with a constant fiber length of $200\,\mu\text{m}$, especially by means of averaged stiffnesses.

The comparison of the effective elastic properties predicted by the different homogenization schemes yields the following results:

- The comparison of FFT-BS and MF-IDD by means of a relative deviation of Young's modulus reveals the following tendencies for all considered microstructures with a fiber volume fraction of 13%: the relative deviation increases with increasing phase contrast and advancing alignment of the fibers. The relative deviation varies between 1.5% and 31.8%.

- In case of UD microstructures, the deviation of FFT-BS and MF-IDD increases with increasing phase contrast and with increasing fiber volume fraction. It varies between 3% and 49.5%.

- The relative deviation between FFT-BS and MF-SC shows another tendency for microstructures with a fiber volume fraction of 13%: It increases with increasing phase contrast but decreases with advancing alignment of the fibers. The relative deviation varies between 4.5% and 68.5%.

- For UD microstructures, the deviation between FFT-BS and MF-SC again increases with increasing phase contrast and with increasing fiber volume fraction. The relative deviation varies between 4.5% and 29.2%.

- If the cylindrical fibers are approximated with ellipsoids not under the conditions of equal aspect-ratio and volume, but, of equal length and volume, the tendencies of the relative deviations between MF and FFT approaches remain the same. However, it affects the comparison of FFT-BS and FFT-IDD in such way, that the deviations become slightly larger. In the comparison of FFT-BS with FFT-SC the deviation become slightly smaller.

Hence, under the assumption of first type ellipsoids in the MF approaches, quantitatively similar results of FFT-IDD and MF-BS approaches, i.e., a maximum relative deviation of 10%, are obtained for all considered microstructure types with a maximum phase contrast of 100 and a maximum fiber volume fraction of 17%. In the case of MF-SC and FFT-BS, only UD microstructures with a maximum phase contrast of 44 deliver a deviation, which suits to this condition. If second type ellipsoids are assumed for MF approaches, the relative deviation between FFT-BS and MF-IDD2 is smaller than 10% only for microstructures with a phase contrast of 44. The same is valid for the comparison of FFT-BS and MF-SC. Additionally, in the last comparison the UD microstructures with $c_{\mathsf{F}} = 13\%$ for all considered phase contrast and $c_F = 17\%$ for $\xi = 100$ deliver small deviations.

The effect of the geometrical fiber modeling assumptions on the effective elastic properties can be summarized as follows:

- Apart from the volume fraction, the geometrical shape of the inclusions significantly affects the effective elastic properties predicted by the mean-field and the full-field approaches. The FFT-BS solution predicts a deviation of the effective Young's modulus in fiber orientation of 5% for the UD microstructures with a fiber volume fraction of 13% if cylindrical fibers are replaced with ellipsoidal approximations of equal aspect ratio.

Hence the shape of the inclusions affect the prediction of the effective elastic properties.

Altogether, it can be concluded, that the selection of the appropriate homogenization method should be done with regard to the problem at hand. The presented mean-field approaches are numerically less expensive than the full-field FFT method. On the one hand, this property allows to handle an equivalent amount of discrete microstructure datasets in shorter time or, on the other hand, to consider more expanded data like segmented microstructure data experimentally measured through micro-computer tomography. The presented FFT-method is able to account for more complex geometrical characteristics, like long curved fibers, than the discussed mean-field approaches. Additionally, the full-field approach is numerically more efficient than other established full-field methods like, e.g., FEM.

5 Prediction of elastic properties using fiber orientation tensors[1]

5.1 Introduction

The mechanical properties of composite materials like SFRCs are crucially dominated by their microstructure. An essential attribute of their microstructure is the orientation distribution of the fibers. The one point statistics of the fiber orientations can be described with fiber orientation distribution functions (FODFs) or, equivalently, with an infinite set of orientation tensors (Torquato, 2002). Kanatani (1984) distinguishes three kinds of orientation tensors, which are also called fabric tensors, orientation-moments or order parameters (see, e.g., Gurp (1998)).

Second-order orientation tensors frequently appear in literature dealing with composite materials (Advani and Tucker III, 1987; Bernasconi et al., 2008, 2012; Dray et al., 2007; Hine et al., 2004; Jördens et al., 2010). Bernasconi et al. (2012), for example, analyzed the fiber orientation distribution of a SFRC using two methods: the

[1] This chapter is based on the paper "Prediction of effective elastic properties of short-fiber reinforced composites using fiber orientation tensors" (Müller and Böhlke, 2015)

first method is based on the observation of the elliptical footprints of the fibers on polished cross sections. The second method starts from a micro-computed tomography scan of the material, which then is examined by an image analysis procedure. The results of both methods are compared by means of the second-order orientation tensor.

Second-order orientation tensors are usually the only microstructure related output variable of mold flow simulation packages (Yang et al., 2010; De Monte et al., 2010). Dray et al. (2007) calculated the thermoelastic properties for an injection molded SFRC. This was done using experimentally determined second and fourth-order orientations tensors, and, additionally, second-order orientation tensors predicted by mold flow simulation. In the latter case, it was necessary to calculate the fourth-order orientation tensor applying closure algorithms. The authors demonstrated the dependence of the thermoelastic properties on the applied closure algorithm (see, e.g., Chung and Kwon (2002)).

Many closure algorithms have been proposed in literature (Advani and Tucker III, 1990; Cintra and Tucker III, 1995; Chung and Kwon, 2002). Some of them, like the linear or the quadratic closure, are directly based on the second-order orientation tensor. Others, like the orthotropic fitted closure, the eigenvalue fitted or the invariant based optimal fitted closures are based on flow data from the calculation of the orientation distribution function. The accuracy of the approximation of the non-fitted closures is only acceptable in particular cases. The fitted closures utilize additional information to calculate their predictions; and still the FODF is not available.

Further, second-order orientation tensors play a role in the homogenization of, e.g., SFRC using full-field approaches. Müller et al.

(2015a) compared predictions of the elastic properties of full-field and mean-field homogenization approaches by means of artificially generated short-fiber microstructure data. The microstructure generation process was based only on the second-order orientation tensor. Hence, the second-order orientation tensor is commonly used and a well established quantity describing approximately the microstructure of certain composites. However, the question remains open, whether the second-order orientation information is sufficient for the prediction of elastic properties. In the present work, this question is addressed for the special case of transversal symmetric fiber orientation distributions.

In this chapter, the properties of FODFs and a classification of different kinds of orientation tensors are described, especially, with a focus on orientation tensors with transversal symmetry. Then, it is shown, how FODF can be estimated if only the elading orientation tensors are available. These FODFs are then used within the homogenization methods SC and IDD in order to calculate the elastic predictions for the properties of PPGF30. Regarding the class of materials with transversal symmetric orientaiton distribution of fibers, it is discussed, whether the second order orientation tensor is a sufficient description of the microstructure in order to predict the elastic properties.

5.2 Fiber orientation distribution function

5.2.1 Properties of FODFs

Throughout this work, the terms "orientation" or "direction" in context of fiber orientations are both describing an axis $\boldsymbol{n}$ of a straight fiber with a constant diameter d and circular profile. The FODF specifies the volume fraction $\mathrm{d}v/v$ of all fibers with a certain orientation $\boldsymbol{n}$ (see, e.g., Zheng and Zou (2001)):

$$\frac{\mathrm{d}v}{v}(\boldsymbol{n}) = f(\boldsymbol{n})\,\mathrm{d}S. \tag{5.1}$$

The quantity $\mathrm{d}S$ is a surface element of the unit sphere $S := \{\boldsymbol{n} \in \mathbb{R}^3 : \|\boldsymbol{n}\| = 1\}$ in the three-dimensional Euclidean space $\mathbb{R}^3$. In spherical coordinates, $\mathrm{d}S = \sin(\vartheta)\,\mathrm{d}\varphi\,\mathrm{d}\vartheta/\,(4\pi)$ holds with the polar and azimuthal angles ϑ and φ.

Generally, an FODF is normalized and non-negative:

$$\int_S f(\boldsymbol{n})\,\mathrm{d}S = 1, \qquad f(\boldsymbol{n}) \geq 0\,\forall \boldsymbol{n} \in S. \tag{5.2}$$

Since fibers are not truly directional, $f(\boldsymbol{n}) = f(-\boldsymbol{n})$ holds. FODFs with this property are called to be centrosymmetric (Zheng and Zou, 2001) or antipodal symmetric (Tyler, 1987).

5.2.2 Empirical FODF

For a set of N equal weighted fiber orientations $\boldsymbol{n}$, the empirical FODF is defined as

$$f(\boldsymbol{n}) = \frac{1}{N}\sum_{\alpha=1}^{N} \delta(\boldsymbol{n} - \boldsymbol{n}_\alpha). \tag{5.3}$$

Herein, $\delta(\boldsymbol{n} - \boldsymbol{n}_\alpha)$ is the Dirac delta distribution.

5.2.3 Orientation tensors

Kanatani (1984) distinguished three different kinds of orientation tensors. The orientation tensors of the first kind are moment tensors of the dyadic products of the direction $\boldsymbol{n}$:

$$\boldsymbol{N}_{\langle\beta\rangle} = \int_S f(\boldsymbol{n})\, \boldsymbol{n}^{\otimes\beta}\, \mathrm{d}S, \tag{5.4}$$

whereas, $\boldsymbol{n}^{\otimes\beta}$ specifies a $(\beta - 1)$-times tensor product. In case of the empirically defined FODF, the first kind orientation tensors result to

$$\boldsymbol{N}_{\langle\beta\rangle} = \frac{1}{N}\sum_{\alpha=1}^{N} \boldsymbol{n}_{\alpha}^{\otimes\beta}. \tag{5.5}$$

Orientation tensors of the first kind are entirely symmetric. Since the FODF is an even function, they are only of even rank. A contraction of a β-order tensor reduces the rank and delivers the orientation tensor $\boldsymbol{N}_{\langle\beta-2\rangle}$:

$$\boldsymbol{N}_{\langle\beta\rangle}\,[\boldsymbol{I}] = \boldsymbol{N}_{\langle\beta-2\rangle}\, \forall\beta \in \{2, 4, 6, \dots\}. \tag{5.6}$$

Hence, the orientation tensors of the first kind are not linearly independent. Since the orientation tensors of the second kind do not play a role in this work, the reader is referred to (Kanatani, 1984) for details.

The orientation tensors of the third kind are defined as the entirely symmetric and traceless part of the first-kind orientation tensors:

$$\boldsymbol{D}_{\langle\alpha\rangle} = \left(\boldsymbol{N}_{\langle\alpha\rangle}\right)'. \tag{5.7}$$

These tensors are referred to as irreducible tensors. It can be shown, that they are linear independent from each other (Kanatani, 1984).

5.2.4 Irreducible orientation tensors of second and fourth-order with transversal symmetry

An irreducible orientation tensor of second-order $\boldsymbol{D}_{\langle 2\rangle}$, which possesses a transversal symmetry with respect to the $\boldsymbol{e}_3$-axis of an orthonormal basis system $\{\boldsymbol{e}_1, \boldsymbol{e}_2, \boldsymbol{e}_3\}$, can be fully described with one independent parameter ξ:

$$\boldsymbol{D}_{\langle 2\rangle} = -\frac{1}{2}\xi\left(\boldsymbol{e}_1 \otimes \boldsymbol{e}_1 + \boldsymbol{e}_2 \otimes \boldsymbol{e}_2\right) + \xi \boldsymbol{e}_3 \otimes \boldsymbol{e}_3. \tag{5.8}$$

An irreducible fourth-order tensor $\boldsymbol{D}_{\langle 4\rangle}$ with a transversal symmetry with respect to the $\boldsymbol{e}_3$-axis can also be fully described with one parameter (Böhlke and Lobos, 2014). Using the normalized Voigt notation, as introduced by Cowin (1989), the components of this tensor can be denoted by a six times six matrix:

$$D^{\langle 4\rangle}_{\alpha\beta} = \begin{bmatrix} 3\psi & \psi & -4\psi & 0 & 0 & 0 \\ \psi & 3\psi & -4\psi & 0 & 0 & 0 \\ -4\psi & -4\psi & 8\psi & 0 & 0 & 0 \\ 0 & 0 & 0 & -8\psi & 0 & 0 \\ 0 & 0 & 0 & 0 & -8\psi & 0 \\ 0 & 0 & 0 & 0 & 0 & 2\psi \end{bmatrix}. \tag{5.9}$$

To represent $\boldsymbol{D}_{\langle 4\rangle} = \sum_{\alpha,\beta=1}^{6} D^{\langle 4\rangle}_{\alpha\beta} \boldsymbol{B}_\alpha \otimes \boldsymbol{B}_\beta$ by a matrix in the last equation, the following basis tensor have been used:

$$\begin{aligned} \boldsymbol{B}_1 &= \boldsymbol{e}_1 \otimes \boldsymbol{e}_1, & \boldsymbol{B}_4 &= \frac{\sqrt{2}}{2}\left(\boldsymbol{e}_2 \otimes \boldsymbol{e}_3 + \boldsymbol{e}_3 \otimes \boldsymbol{e}_2\right), \\ \boldsymbol{B}_2 &= \boldsymbol{e}_2 \otimes \boldsymbol{e}_2, & \boldsymbol{B}_5 &= \frac{\sqrt{2}}{2}\left(\boldsymbol{e}_1 \otimes \boldsymbol{e}_3 + \boldsymbol{e}_3 \otimes \boldsymbol{e}_1\right), \\ \boldsymbol{B}_3 &= \boldsymbol{e}_3 \otimes \boldsymbol{e}_3, & \boldsymbol{B}_6 &= \frac{\sqrt{2}}{2}\left(\boldsymbol{e}_1 \otimes \boldsymbol{e}_2 + \boldsymbol{e}_2 \otimes \boldsymbol{e}_1\right). \end{aligned} \tag{5.10}$$

Both irreducible tensors are included in the fourth-order orientation tensor of the first kind $\boldsymbol{N}_{\langle 4\rangle}$.

Since a transversally isotropic $\boldsymbol{N}_{\langle 4\rangle}$ is entirely symmetric and $\mathrm{sp}(\boldsymbol{N}_{\langle 2\rangle}) = \mathrm{sp}(\boldsymbol{N}_{\langle 4\rangle}[\boldsymbol{I}]) = 1$ holds, this tensor depends on two parameters:

$$N^{\langle 4\rangle}_{\alpha\beta} = \begin{bmatrix} 3\gamma & \gamma & \delta & 0 & 0 & 0 \\ \gamma & 3\gamma & \delta & 0 & 0 & 0 \\ \delta & \delta & 1-8\gamma-4\delta & 0 & 0 & 0 \\ 0 & 0 & 0 & 2\delta & 0 & 0 \\ 0 & 0 & 0 & 0 & 2\delta & 0 \\ 0 & 0 & 0 & 0 & 0 & 2\gamma \end{bmatrix}, \tag{5.11}$$

with $\gamma = (7 - 15\xi + 105\psi)/105$ and $\delta = (14 + 15\xi - 840\psi)/210$.

For valid orientation tensors, ξ and ψ can not be chosen in a completely arbitrary way. According to its definition in equation (5.4) or (5.5), fourth-order orientation tensors of the first-kind are entirely symmetric and positive semi-definite with respect to an arbitrary second-order tensor $\boldsymbol{M}_{\langle 2\rangle}$:

$$\boldsymbol{M}_{\langle 2\rangle} \cdot \boldsymbol{N}_{\langle 4\rangle}[\boldsymbol{M}_{\langle 2\rangle}] \geq 0. \tag{5.12}$$

The solution of this eigenvalue problem yields the range for the parameters ξ and ψ (see Figure 5.1 and Nomura et al. (1970)):

$$-\frac{1}{3} \leq \xi \leq \frac{2}{3}, \qquad -\frac{1}{90} - \frac{1}{42}\xi + \frac{1}{8}\xi^2 \leq \psi \leq \frac{1}{60} + \frac{1}{56}\xi. \tag{5.13}$$

The space bounded by this inequalities is shown in Figure 5.1.

5.3 Estimation of the FODF based on leading orientation tensors

5.3.1 Information-theoretic entropy

The inherent incompleteness of measured data implies mathematical problems, which do not have unique solutions. Such problems are called ill-posed. With the *maximum entropy principle* (MEP), it is possible to tackle ill-posed problems and to single out one solution by choosing the solution with the maximal entropy. Initially, Shannon (1948) identified a quantity in the context of information theory, that is a measure of uncertainty of an information source. Due to the related meaning and the equivalent mathematical formulation of this quantity to the entropy in thermodynamics, it is called *Shannon's-entropy* or *information-theoretic entropy*.

Jaynes (1957a,b) introduced the MEP in the field of statistical mechanics. Since then, MEP has been applied in, e.g., spectral analysis (Ulrych and Bishop, 1975) or image restoration (Frieden, 1972). MEP has been applied to solve differential equations (Mead, 1986), moment problems in texture analysis (Böhlke, 2005), and in the simulation of crystallographic textures (Böhlke, 2006). Junk et al. (2012) have shown the existence of solutions of moment problems in texture analysis for a specific crystal symmetry. A comprehensive overview of the MEP is given in Wu (1997).

The information-theoretic entropy E is defined by

$$E = -\int_S f(\boldsymbol{n}) \ln \left(f(\boldsymbol{n}) \right) \mathrm{d}S \; \in (-\infty, 0], \qquad (5.14)$$

where $f(\boldsymbol{n})$ is an FODF. In case of a uniform distribution, $E = 0$ holds.

5.3.2 Moment problem

The approximation of an FODF based on the leading orientation tensors only, can be formulated as a moment problem. Since the moment of the function $\boldsymbol{n}^{\otimes\alpha}$ regarding $f(\boldsymbol{n})$ is computed by (5.4), orientation tensors $\boldsymbol{N}_{\langle\alpha\rangle}$ can be considered as averages or expectations of the corresponding functions $\boldsymbol{n}^{\otimes\alpha}$. Thus, in order to consistently estimate $f(\boldsymbol{n})$ by $\bar{f}(\boldsymbol{n})$ using N given irreducible orientation tensors $\{\boldsymbol{D}_{\langle 2\rangle}, \ldots, \boldsymbol{D}_{\langle N\rangle}\}$, the moment problem is stated as follows:

$$\begin{aligned} \bar{E} &= -\int_S \bar{f}(\boldsymbol{n}) \ln\left(\bar{f}(\boldsymbol{n})\right) \mathrm{d}S \quad \rightarrow \quad \max, \\ \boldsymbol{G}_{\langle 0\rangle} &:= \int_S \bar{f}(\boldsymbol{n})\, \mathrm{d}S - 1 \overset{!}{=} 0, \\ \boldsymbol{G}_{\langle\alpha\rangle} &:= \int_S \bar{f}(\boldsymbol{n}) \left(\boldsymbol{n}^{\otimes\alpha}\right)' \mathrm{d}S - \boldsymbol{D}_{\langle\alpha\rangle} \overset{!}{=} 0, \end{aligned} \tag{5.15}$$

where α is even and ranges from 2 to N.

The first condition $\boldsymbol{G}_{\langle 0\rangle}$ represents the normalization condition in equation (5.2). The additional conditions take into account the given orientation tensors to be reproduced by the function $\bar{f}(\boldsymbol{n})$.

Since an irreducible tensor $\boldsymbol{D}_{\langle\alpha\rangle}$ of rank α has $2\alpha + 1$ independent coefficients, there is a number of $1 + \sum_{\alpha=2}^{N}(2\alpha + 1)$ equations to be considered within the constrained maximization problem. Using the Lagrange multiplier method, this problem can be solved with the following Lagrange functional:

$$\mathcal{L} = \bar{E} - \sum_{\alpha=0}^{N} \boldsymbol{L}_{\langle\alpha\rangle} \cdot \boldsymbol{G}_{\langle\alpha\rangle} \quad \forall\alpha \in \{0, 2, 4, \dots\}. \tag{5.16}$$

Herein, $\boldsymbol{L}_{\langle\alpha\rangle}$ are the Lagrange multipliers. The first variation of $\mathcal{L}$ with respect to $\bar{f}(\boldsymbol{n})$ has to vanish:

$$\delta\mathcal{L} = 0 = \int_S q(\boldsymbol{n})\delta\bar{f}(\boldsymbol{n})\,\mathrm{d}S, \tag{5.17}$$

with $q(\boldsymbol{n}) = -1 - \ln\left(\bar{f}(\boldsymbol{n})\right) - \boldsymbol{L}_0 - \sum_{\alpha=2}^{N} \boldsymbol{L}_{\langle\alpha\rangle} \cdot (\boldsymbol{n}^{\otimes\alpha})'$. For arbitrary $\delta\bar{f}(\boldsymbol{n})$, the function $q(\boldsymbol{n})$ has to be zero. Thus, the approximating density function results to

$$\bar{f}(\boldsymbol{n}) = \exp\left(-1 - \boldsymbol{L}_0 - \sum_{\alpha=2}^{N} \boldsymbol{L}_{\langle\alpha\rangle} \cdot \left(\boldsymbol{n}^{\otimes\alpha}\right)'\right) \quad \forall\alpha \in \{2, 4, \dots\}. \tag{5.18}$$

Numerically, this problem has been solved by inserting $\bar{f}(\boldsymbol{n})$ in the conditions $\boldsymbol{G}_{\langle 0\rangle}$ and $\boldsymbol{G}_{\langle\alpha\rangle}$ and calculating the Lagrange multipliers using a damped Newton-Raphson algorithm. The integration has been carried out by an adaptive quadrature scheme (Hahn, 2005).

5.4 Homogenization of elastic properties using FODFs

5.4.1 Self-consistent homogenization

The first mean-field homogenization method applied is the self-consistent scheme based on the formulation of Willis (1981). According to this, the effective elastic stiffness tensor $\mathbb{C}^{\mathrm{SC}}$ for a $N+1$-phase particulate composite is given by

$$\mathbb{C}^{\mathrm{SC}} = \mathbb{C}_{\mathsf{M}} + \sum_{\alpha=1}^{N} c_\alpha \left(\mathbb{C}_\alpha - \mathbb{C}_{\mathsf{M}}\right) \left(\mathbb{I}^{\mathrm{s}} + \mathbb{P}_\alpha^{\mathrm{SC}} \left(\mathbb{C}_\alpha - \mathbb{C}^{\mathrm{SC}}\right)\right)^{-1}, \tag{5.19}$$

with c_α and $\mathbb{C}_\alpha$ being the volume fraction and stiffness tensor of phase α. N corresponds to the number of phases inside the considered volume element. The index $(\cdot)_{N+1}$ is replaced by $(\cdot)_{\mathsf{M}}$, which denotes the matrix phase. $\mathbb{P}_\alpha^{\mathrm{SC}}$ denotes Hill's polarization tensor, see equation (3.9). Details on the self-consistent approach in terms of a discrete microstructure description are given in section 3.4.2.

Equation (5.19) is valid for a matrix-inclusion composite with N inclusions. A similar prescription for the effective stiffness is found, if a composite with $N \to \infty$ phases with a total volume fraction c_{F} is considered, where the phases only differ in terms of their orientation, i.e., uniform aspect-ratio and stiffness:

$$\mathbb{C}^{\mathrm{SC}} = \mathbb{C}_{\mathsf{M}} + c_{\mathsf{F}} \int_S f(\boldsymbol{n}) \left(\mathbb{C}(\boldsymbol{n}) - \mathbb{C}_{\mathsf{M}}\right) \mathbb{A}(\boldsymbol{n}) \, \mathrm{d}S, \tag{5.20}$$

with $\mathbb{A}(\boldsymbol{n}) = \left(\mathbb{I}^{\mathrm{s}} + \mathbb{P}^{\mathrm{SC}}(\boldsymbol{n}) \left(\mathbb{C}(\boldsymbol{n}) - \mathbb{C}^{\mathrm{SC}}\right)\right)^{-1}$, the FODF $f(\boldsymbol{n})$, and the orientation dependent stiffness $\mathbb{C}(\boldsymbol{n})$. The polarization tensor $\mathbb{P}^{\mathrm{SC}}$ still depends on $\mathbb{C}^{\mathrm{SC}}$ and the description of the ellipsoidal geometry $\boldsymbol{Z}(\boldsymbol{n}) = \boldsymbol{Q}^{\mathsf{T}} \boldsymbol{Z}_0 \boldsymbol{Q}$. Whereby $\boldsymbol{Q} = \boldsymbol{Q}(\boldsymbol{n})$ is an orthogonal tensor, which rotates the reference geometry $\boldsymbol{Z}_0$ in the direction $\boldsymbol{n}$.

5.4.2 Interaction direct derivative estimate

The second mean-field method applied is the interaction direct derivate (IDD) estimate, see section 3.4.3. The advantage of the IDD lies in its explicit structure, which is valid for multi-phase composites with different material symmetries and distributions.

Similar to equation (5.20), the following prescription gives the IDD estimate of the properties of the effective medium for an ∞-phase composite, where the inclusion phases amount to a total volume fraction of c_{F}, and differ only in their orientation:

$$\mathbb{C}^{\mathrm{IDD}} = \mathbb{C}_{\mathsf{M}} + \left(\mathbb{I}^{\mathrm{s}} - c_{\mathsf{F}} \int_S f(\boldsymbol{n}) \mathbb{N}(\boldsymbol{n}) \mathbb{P}^{\mathsf{D}}(\boldsymbol{n}) \, \mathrm{d}S \right)^{-1} \left(c_{\mathsf{F}} \int_S f(\boldsymbol{n}) \mathbb{N}(\boldsymbol{n}) \, \mathrm{d}S \right), \tag{5.21}$$

with $\mathbb{N}(\boldsymbol{n}) = (\mathbb{C}(\boldsymbol{n}) - \mathbb{C}_{\mathsf{M}}) \left(\mathbb{I}^{\mathrm{s}} + \mathbb{P}(\boldsymbol{n}) \left(\mathbb{C}(\boldsymbol{n}) - \mathbb{C}_{\mathsf{M}} \right) \right)^{-1}$.

Here, $\mathbb{P} = \mathbb{P}(\mathbb{C}_{\mathsf{M}}, \boldsymbol{Z}(\boldsymbol{n}))$ is the Hill's polarization tensor as defined in equation (3.9), however, $\mathbb{C}^{\mathsf{SC}}$ is replaced with $\mathbb{C}_{\mathsf{M}}$. If the matrix-inclusion cell takes on an ellipsoidal shape, then $\mathbb{P}^{\mathsf{D}} = \mathbb{P}(\mathbb{C}_{\mathsf{M}}, \boldsymbol{Z}^{\mathsf{D}}(\boldsymbol{n}))$ is the polarization tensor corresponding to an ellipsoidal inclusion with geometry of the matrix-inclusion cell $\boldsymbol{Z}^{\mathsf{D}}(\boldsymbol{n})$, which is embedded in an infinite matrix with the stiffness $\mathbb{C}_{\mathsf{M}}$. The shape of the matrix-inclusion cell defines the inclusion distribution in the composite. In the present work, $\boldsymbol{Z}^{\mathsf{D}}(\boldsymbol{n}) = \boldsymbol{Z}(\boldsymbol{n})$ is assumed.

5.5 Results

5.5.1 Model parameter

Exemplarily, the properties of the composite material PPGF30 are considered. Both constituents, the polypropylene matrix ($E_{\mathsf{M}} = 1.705\,\mathrm{GPa}$, $\nu_{\mathsf{M}} = 0.355$) and the glass fibers ($E_{\mathsf{F}} = 73.0\,\mathrm{GPa}$, $\nu_{\mathsf{F}} = 0.22$) are assumed to be linear elastic and isotropic. The microstructure of PPGF30 is modeled by a constant aspect ratio

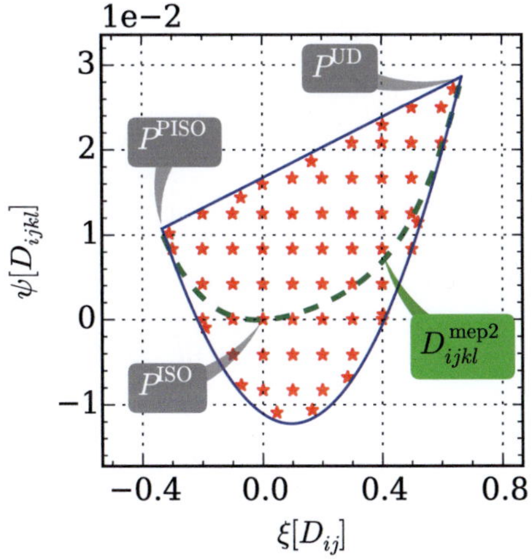

Figure 5.1: Space for the irreducible parameters ξ and ψ, and its discretization by pairs of ξ and ψ values

of $a = 20$ for the fibers, and a fiber orientation distribution with transversal symmetry given by an FODF $f(\boldsymbol{n})$. The FODF is defined in terms of the parameters ξ and ψ (see section 5.2.4). The discretization of the space for valid orientation tensors is shown in Figure 5.1. The parameter combinations $(\xi = 2/3, \psi = 1/35)$, $(\xi = -1/3, \psi = 3/280)$ and $(\xi = 0.0, \psi = 0.0)$ denote the unidirectional (P^{UD}), the planar isotropic (P^{PISO}) and the isotropic (P^{ISO}) distribution, respectively. The FODF is calculated in two ways: firstly, only the second-order orientation tensor, described by the parameter ξ, is assumed to be given. Secondly, the second and fourth-order orientation tensors are assumed to be given. All results based on the first and second approximations are marked with $(\cdot)^{\mathsf{mep2}}$ and $(\cdot)^{\mathsf{mep4}}$, respectively, in the following. The estimate of the fourth-order tensor based on the FODF f^{mep2} is indicated by a green dashed line in the Figure 5.1, and in the following figures, also.

5.5.2 Anisotropy of estimated FODFs

The deviation of an FODF from a uniform distribution can be interpreted as the anisotropy of this FODF. The following measure is applied, in order to quantify this anisotropy and to compare the FODFs with each other:

$$K(f_1, f_2) = -\int_S f_1(\boldsymbol{n}) \ln\left(\frac{f_1(\boldsymbol{n})}{f_2(\boldsymbol{n})}\right) \mathrm{d}S \in (-\infty, 0]. \tag{5.22}$$

This measure is referred to as the *entropy* or the *Kullback information* (Kullback, 1968; Kanatani, 1984). It vanishes, if the functions are equal, and tends to minus infinity for independent functions $f_1(\boldsymbol{n})$ and $f_2(\boldsymbol{n})$.

In Figure 5.2, the anisotropy of f^{mep2}, f^{mep4} and a comparison of these two functions is shown. f^{mep2} depends only on ξ, see Figure 5.2(a). Thus, starting at P^{ISO}, the anisotropy of f^{mep2} increases with increasing and decreasing ξ. At P^{UD}, the anisotropy reaches its peak. Since f^{mep4} depends on both parameters, the anisotropy increases with increasing distance from P^{ISO}, see Figure 5.2(b). Again, higher anisotropy can be observed at the boundaries of the region and the peak at P^{UD}. The comparison of f^{mep2} and f^{mep4} in Figure 5.2(c) reveals a fairly good conformity in the center of the region, at P^{PISO} and P^{UD}. At the bounds, a greater deviation can be observed.

5.5.3 Comparison of effective elastic properties

For both approximations, f^{mep2} and f^{mep4}, the SC and the IDD method have been applied to calculate the effective elastic stiffness. The results have been compared by means of three quantities:

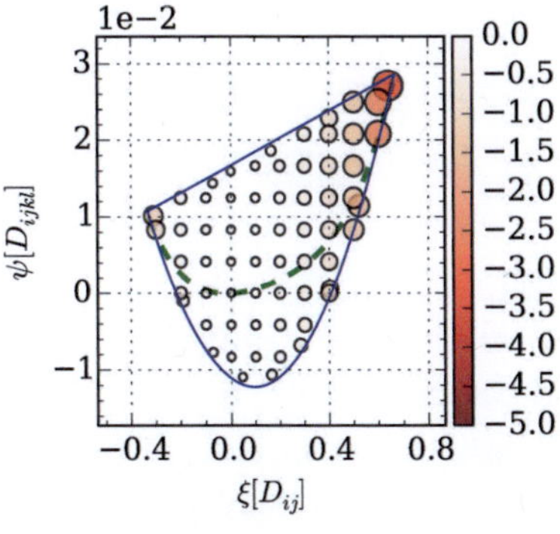

(a) $K\left(f^{\mathrm{mep2}}, 1\right)$

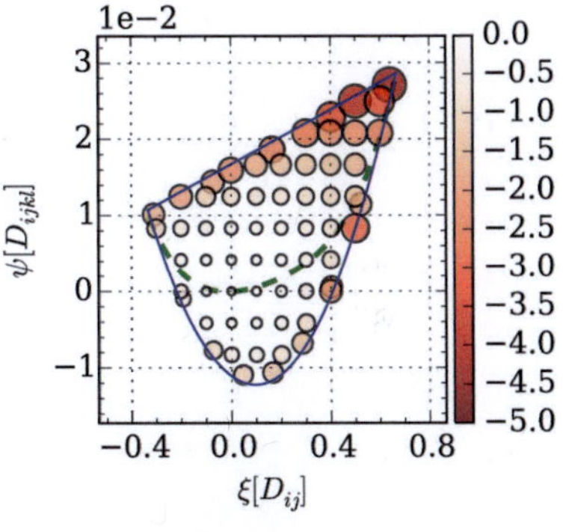

(b) $K\left(f^{\mathrm{mep4}}, 1\right)$

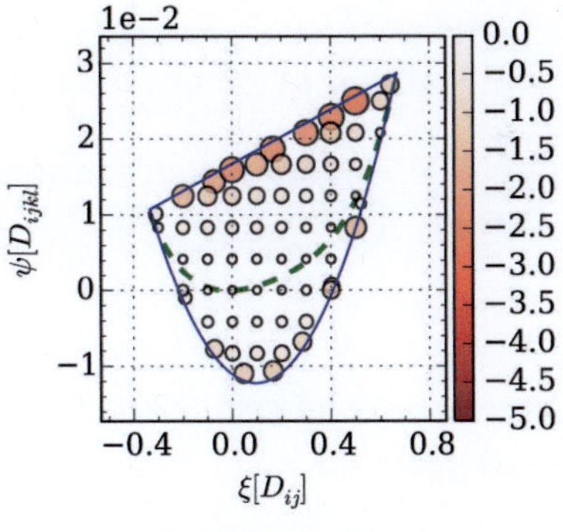

(c) $K\left(f^{\mathrm{mep4}}, f^{\mathrm{mep2}}\right)$

Figure 5.2: Anisotropy and comparison of the estimated FODFs f^{mep2} and f^{mep4}

firstly, the anisotropy of the effective stiffness tensors have been calculated:

$$A(\mathbb{C}) = \frac{\lambda_{\min}\left(\mathbb{H}(\mathbb{C})\right)}{\lambda_{\max}\left(\mathbb{H}(\mathbb{C})\right)} \in (0,1], \tag{5.23}$$

with $\mathbb{H}(\mathbb{C}) = \exp\left(\mathrm{aniso}\left(\ln\left(\mathbb{C}\right)\right)\right)$. The function $\mathrm{aniso}\left(\mathbb{C}\right)$ is defined as

$$\mathrm{aniso}\left(\mathbb{C}\right) = \mathbb{C} - \sum_{\alpha=1}^{2}\left(\mathbb{C}\cdot\frac{\mathbb{P}_\alpha}{||\mathbb{P}_\alpha||}\right)\frac{\mathbb{P}_\alpha}{||\mathbb{P}_\alpha||}, \tag{5.24}$$

with the first and second isotropic projector $\mathbb{P}_1 = \frac{1}{3}\boldsymbol{I}\otimes\boldsymbol{I}$ and $\mathbb{P}_2 = \mathbb{I}^{\mathrm{s}} - \mathbb{P}_1$. This anisotropy measure delivers the same dimensionless scalar value for a stiffness and the corresponding compliance tensor. The $||\cdot||$-function is the Frobenius norm. Secondly, the stiffness tensors have been compared directly using the following normalized measure of two real tensorial or non-tensorial variables C and $\tilde{C}$:

$$\eta_C\left(C,\tilde{C}\right) = \frac{||C-\tilde{C}||}{||C||}. \tag{5.25}$$

Thirdly, the directional dependent Young's modulus $E(\boldsymbol{n})$

$$\frac{1}{E\left(\boldsymbol{n}\right)} = \boldsymbol{n}\otimes\boldsymbol{n}\cdot\mathbb{S}[\boldsymbol{n}\otimes\boldsymbol{n}], \tag{5.26}$$

with the compliance tensor $\mathbb{S} = \mathbb{C}^{-1}$, has been used for comparisons by means of a relative deviation

$$\eta_E(E_1,E_2) = \left(\frac{\int_S\left(E_1\left(\boldsymbol{n}\right) - E_2\left(\boldsymbol{n}\right)\right)^2\,\mathrm{d}S}{\int_S\left(E_1\left(\boldsymbol{n}\right)\right)^2\,\mathrm{d}S}\right)^{1/2}, \tag{5.27}$$

for direct comparisons in particular directions and, finally, for a comparison of its transversal symmetric shape.

Figures 5.3, 5.4 and 5.5 show the SC results. Figures 5.3(a) and 5.3(c)

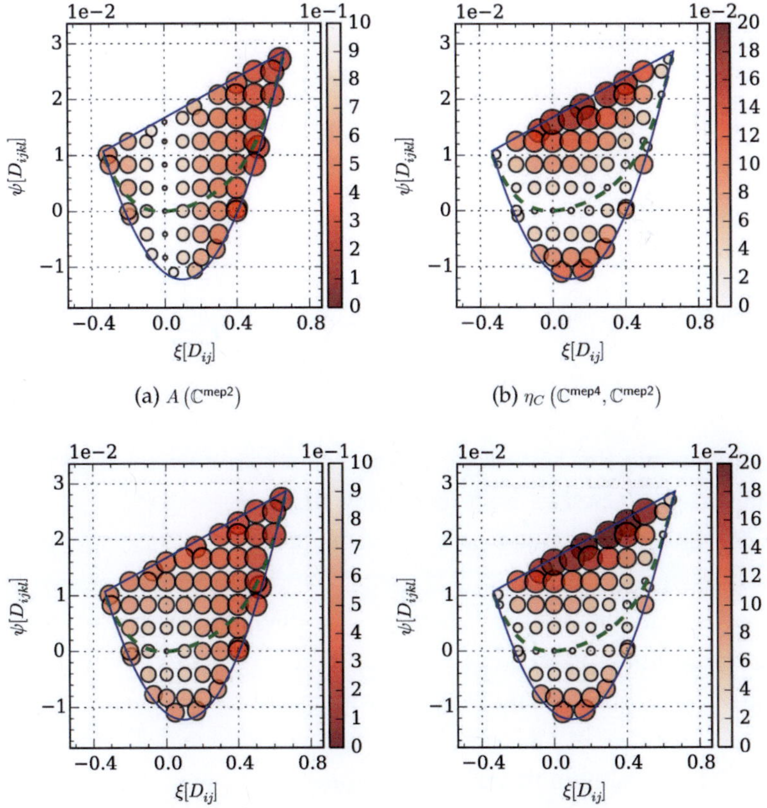

(a) $A\left(\mathbb{C}^{\text{mep2}}\right)$

(b) $\eta_C\left(\mathbb{C}^{\text{mep4}}, \mathbb{C}^{\text{mep2}}\right)$

(c) $A\left(\mathbb{C}^{\text{mep4}}\right)$

(d) $\eta_E\left(E^{\text{mep4}}, E^{\text{mep2}}\right)$

Figure 5.3: Comparison of $(\cdot)^{\text{mep2}}$ and $(\cdot)^{\text{mep4}}$ results for SC homogenization

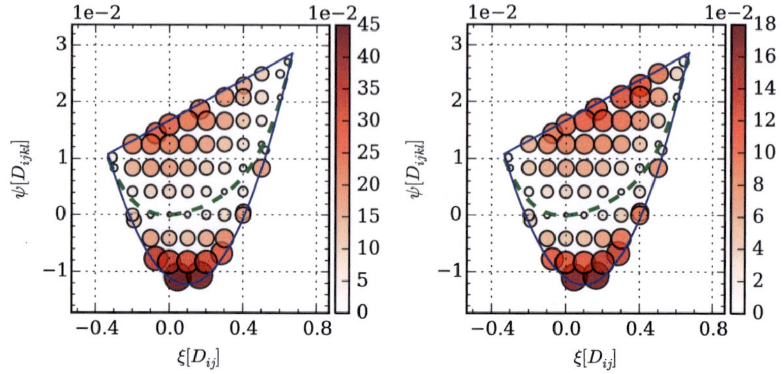

(a) $\eta_C\left(E_0^{\mathrm{mep4}}, E_0^{\mathrm{mep2}}\right)$ (b) $\eta_C\left(E_{90}^{\mathrm{mep4}}, E_{90}^{\mathrm{mep2}}\right)$

Figure 5.4: Comparison of E^{mep2} and E^{mep4} in the direction of the transversal symmetry axis and perpendicular to it for SC homogenization

demonstrate the anisotropy of the effective stiffness. Again, the anisotropy increases with increasing distance from P^{ISO}, whereas $\mathbb{C}^{\mathrm{mep2}}$ does not depend on ψ. The highest anisotropy can be observed at P^{UD}. In Figure 5.3(b), $\mathbb{C}^{\mathrm{mep2}}$ and $\mathbb{C}^{\mathrm{mep4}}$ are compared by means of the relative deviation η_C. The deviation is approximately constant if ψ is fixed and ξ is varied. The largest deviations can be observed at the top and bottom boundary of the parameter region. At P^{UD} and P^{PISO}, the predictions agree fairly well. The maximal deviation amounts to 18% for the considered parameter combinations. Very similar trends can be seen in the figures 5.3(d), 5.4(a) and (b). In Figure 5.3(d), the relative deviation η_E of Young's moduli is shown. For the Figures 5.4(a) and (b), the directional dependent Young's modulus has been evaluated in the direction of the transversal symmetry axis (0°-direction) and perpendicular to this axis (90°-direction), respectively, for $\mathbb{C}^{\mathrm{mep2}}$ and $\mathbb{C}^{\mathrm{mep4}}$. The resulting Young's moduli are then compared by means of η_C. While, the trends are similar, the deviations are very different. The deviation $\eta_E\left(E^{\text{mep4}}, E^{\text{mep2}}\right)$ shows slightly higher values compared to the comparison of the stiffnesses $\eta_C\left(\mathbb{C}^{\text{mep4}}, \mathbb{C}^{\text{mep2}}\right)$. In both cases, the highest deviations occur at the upper bound (20% for η_E and 18% for η_C). However, in $0°$-direction, the deviations are much higher. The largest deviations occur at the lower bound at smaller ψ (46%). The same trend is observed for Young's moduli in $90°$-direction, with a maximal deviation of 17%.

In Figure 5.5, shapes of the directional dependent Young's moduli for particular parameter combinations are shown. In all illustrations, the horizontal axis agrees with the transversal symmetry axis. Due to the antipodal symmetry of the directional dependent Young's modulus, see equation (5.26), the vertical axis is also a symmetry axis. Thus, the full Young's modulus body can be obtained by mirroring the shown shape on the vertical axis and rotating the obtained shape about the horizontal axis. First, in the figures 5.5(a)-(d), ψ is varied, while $\xi = 0.0$ is kept constant. This implies for all cases an isotropic second-order orientation tensor. In this series of figures, E^{mep2} takes on always the same spherical shape. Whereas, E^{mep4} changes its shape, significantly. The curves shown in Figure 5.5(c) denote an isotropic Young's modulus for both approximations. Apart from that configuration, the difference in the shape is distinct. In the second series of figures (5.5(e)-(h)), ξ is varied, while $\psi = 0$ is constant. In this comparison, no distinct difference can be observed.

It is known from literature, that the applied SC approach generally overestimates the true stiffness of a composite with stiff inclusions in a soft matrix. However, the IDD method yields a lower bound for the elastic properties if all inclusions are modeled with the same geometry and distribution (Zheng and Du, 2001). Hence, it is rea-

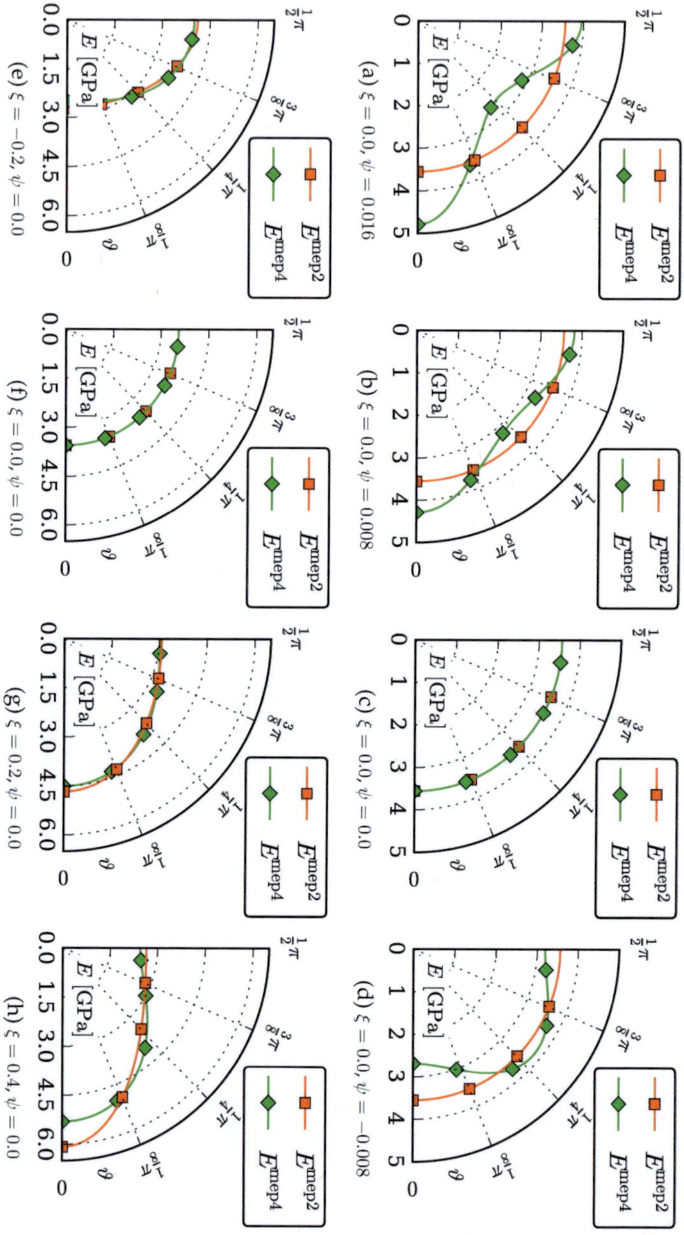

(a) $\xi = 0.0, \psi = 0.016$
(b) $\xi = 0.0, \psi = 0.008$
(c) $\xi = 0.0, \psi = 0.0$
(d) $\xi = 0.0, \psi = -0.008$
(e) $\xi = -0.2, \psi = 0.0$
(f) $\xi = 0.0, \psi = 0.0$
(g) $\xi = 0.2, \psi = 0.0$
(h) $\xi = 0.4, \psi = 0.0$

Figure 5.5: Shape of Young's modulus for different parameter combinations (SC) sonable to compare both methods and the effects of the approximations of FODFs in combination with them. In the figures 5.6 and 5.7, the IDD results are aggregated. In all examples, the anisotropy of the stiffnesses and all deviation measures show the same trends, but they are less pronounced compared to the SC examples. The deviation $\eta_C\left(\mathbb{C}^{\text{mep4}}, \mathbb{C}^{\text{mep2}}\right)$ and $\eta_E\left(E^{\text{mep4}}, E^{\text{mep2}}\right)$, however, still amount to approximately 15.8% and 16.2%, respectively. The deviation of Young's modulus in $0°$ and $90°$-direction comes to 39% and 14%. Thus, the IDD results confirm the observations made above: for constant ψ and variable ξ, the deviations do not change significantly. If ξ is kept constant and ψ is varied, significant changes occur in terms of each considered deviation. The shapes of Young's moduli in terms of IDD are qualitatively equal to that of SC and will, thus, not be examined further.

5.6 Summary and conclusions

The main objective of this chapter is to give an answer to the question: is it sufficient to consider only the second-order fiber orientation tensor as the main quantity describing the microstructure of short-fiber reinforced composites (SFRCs) with transversal isotropic symmetry, especially in terms of the prediction of effective elastic properties?

This topic has been elaborated by means of the SFRC PPGF30. This material consists of a polypropylene matrix and glass fibers with a mass fraction of 30%. The microstructure of this composite is supposed to be governed by a transversal orientation distribution of the fibers with a constant aspect-ratio of 20. The transversal orientation distribution is described by irreducible orientation tensors of

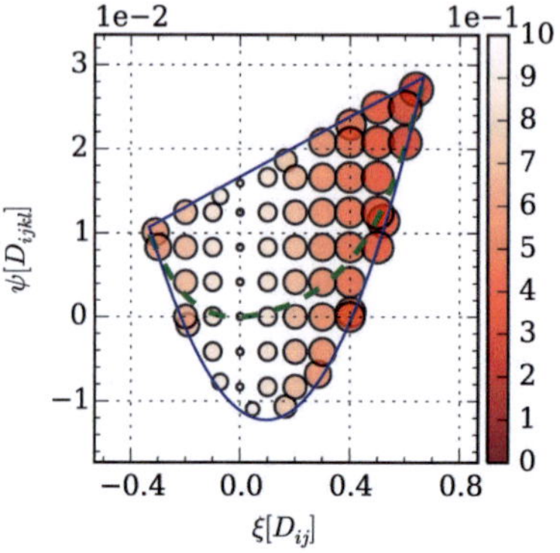

(a) $A\left(\mathbb{C}^{\mathsf{mep2}}\right)$

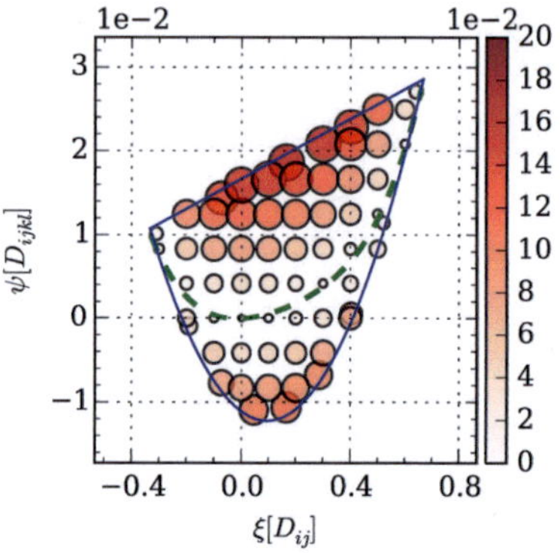

(b) $\eta_C\left(\mathbb{C}^{\mathsf{mep4}}, \mathbb{C}^{\mathsf{mep2}}\right)$

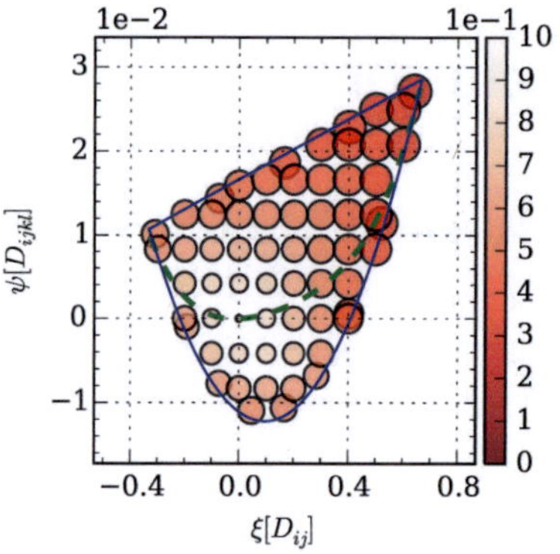

(c) $A\left(\mathbb{C}^{\mathsf{mep4}}\right)$

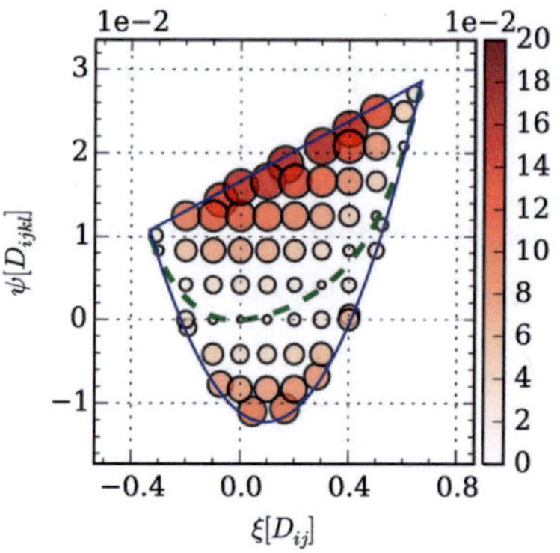

(d) $\eta_E\left(E^{\mathsf{mep4}}, E^{\mathsf{mep2}}\right)$

Figure 5.6: Comparison of $(\cdot)^{\mathsf{mep2}}$ and $(\cdot)^{\mathsf{mep4}}$ results for IDD homogenization

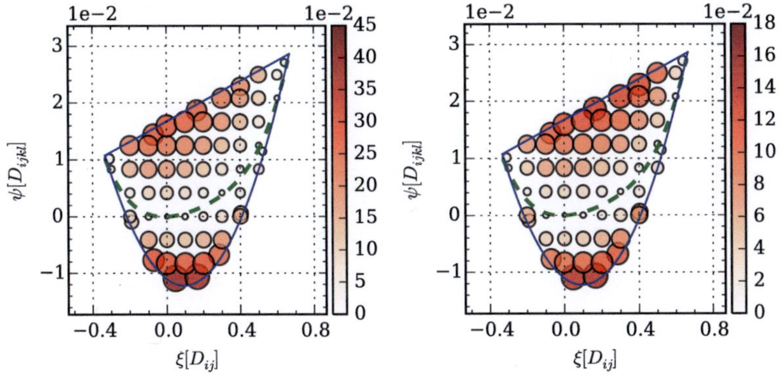

(a) $\eta_C\left(E_0^{\text{mep4}}, E_0^{\text{mep2}}\right)$ (b) $\eta_C\left(E_{90}^{\text{mep4}}, E_{90}^{\text{mep2}}\right)$

Figure 5.7: Comparison of E^{mep2} and E^{mep4} in the direction of the transversal symmetry axis and perpendicular to it for IDD homogenization

second and fourth-order. FODFs have been calculated on the basis of the second-order irreducible orientation tensor, on the one hand, and on the basis of both tensors, on the other hand. The estimation of the FODFs has been taken out within the concept of maximal entropy. The effective elastic properties have been calculated applying the SC and IDD approaches.

Referring to the resulting FODFs f^{mep2} and f^{mep4}, the effective stiffness tensors $\mathbb{C}^{\text{mep2}}$ and $\mathbb{C}^{\text{mep4}}$, and the directional dependent Young's moduli $E^{\text{mep2}}(\boldsymbol{n})$ and $E^{\text{mep4}}(\boldsymbol{n})$, the following conclusions can be made:

- **Properties of the fourth-order orientation tensor of first kind:** By definition, this tensor is entirely symmetric, normalized, and positive semi-definite. For a valid fourth-order orientation tensor, these properties have to be taken into account by closure algorithms

- **Implication of MEP on the fourth-order orientation tensor:** The application of MEP in order to calculate the fourth-order orientation tensor implies a special coupling of ψ and ξ. These couplings are unique for the isotropic point P^{ISO} and for all pairs with $\psi > 3/280$. In all other couplings, ψ appears twice. However, many other couplings are also generally possible which show significant differences in the mechanical properties compared to the MEP estimate.

- **Anisotropy of f^{mep2} and f^{mep4}:** The anisotropy of the FODFs increases with increasing distance from the isotropic point in the ξ-ψ-parameter space of the irreducible second and fourth-order tensors. The difference between f^{mep2} and f^{mep4} is approximately constant if ξ is varied, only. The largest deviations occur at extremal ψ-values. At P^{UD} and P^{PISO}, the FODF approximations are similar.

- **Anisotropy of $\mathbb{C}^{\text{mep2}}$ and $\mathbb{C}^{\text{mep4}}$:** The stiffness tensors are directly affected by the FODFs. Thus, the anisotropy of the stiffness tensor shows the same trends as the corresponding FODF.

- **Comparison of $\mathbb{C}^{\text{mep2}}$ vs $\mathbb{C}^{\text{mep4}}$ and E^{mep2} vs E^{mep4}:** For a constant ψ but variable ξ, the deviation between $\mathbb{C}^{\text{mep2}}$ and $\mathbb{C}^{\text{mep4}}$ in terms of the relative norm $\eta_C(\mathbb{C}^{\text{mep4}}, \mathbb{C}^{\text{mep2}})$ is approximately constant. For a constant ξ, the deviation varies significantly. These observations hold also for the relative deviation of the directional dependent Young's modulus $\eta_E(E^{\text{mep4}}, E^{\text{mep2}})$. The maximal deviations for the stiffnesses and Young's moduli amount to 18% and 20% in case of SC, and 15.8% and 16.2% in case of IDD. Hence, if the second-order orientation tensor is available, only, one has to be aware of a relative error of at least 16%.

- **Young's moduli in particular directions**: In the 0°-direction, which corresponds to the axis of transversal symmetry of the FODF, the deviations partially have very high values: up to 46% in case of SC, and 39% in case of IDD. In the perpendicular 90°-direction, the deviations are smaller: up to 17% in case of SC, and 14% in case of IDD. Hence, relying on the second-order orientation tensor, only, at least an error of 39% may occur in the direction of the transversal symmetry axis.

- **Directional dependence of Young's modulus**: The shapes of Young's moduli for the considered configurations of ξ and ψ confirm the foregoing observations: If ψ is kept constant, the shape of $E^{\mathrm{mep2}}(\boldsymbol{n})$ does not differ significantly from $E^{\mathrm{mep4}}(\boldsymbol{n})$. If ξ is kept constant, significant differences in terms of the shapes of $E^{\mathrm{mep2}}(\boldsymbol{n})$ and $E^{\mathrm{mep4}}(\boldsymbol{n})$ become obvious.

In view of these large deviations, we conclude, that a description of the microstructure of a composite with transversally isotropic symmetry using only the second-order orientation tensor is generally insufficient. However, Hine et al. (2004) concluded that the microstructure of real SFRC follows the maximum entropy statistics. This conclusion was based on fiber orientations determined by image analysis of cross sections measuring the ellipticity of each fiber and, therefore, refer to specific experimental data. Since our first maximum entropy approximated FODF $f^{\mathrm{mep2}}(\boldsymbol{n})$ relies on the second-order orientation tensor, only, it agrees exactly with the transversally isotropic structures of Hine et al. (2004). Thus, if real microstructures indeed follow the maximum entropy statistics, then, in combination with the maximum entropy approach, the second-order orientation tensor is a sufficient description of the microstructure of SFRC in order to predict the elastic properties.

6 Summary and conclusions

Modern fiber reinforced composite offer so many opportunities in designing structural parts and visible surfaces, as well. The possibility to adjust the required properties by combining different materials allows the usage of composites in nearby every conceivable application.

All these possibilities and opportunities comprise, however, a similar amount of challenges. With the aim to design a reliable product, these challenges have to be met. Considering the entire complexity, a serious attempt to lift the level of knowledge a little bit, has to focus on one specific topic.

Hence, the focus of the present work lies on the prediction of elastic behavior of short-fiber reinforced composites. In particular, mean-field homogenization methods are considered, which account for detailed experimentally determined and also artificially constructed microstructure data in a discrete and averaged form.

This set of issues is elaborated by the following three main topics of this work: the first topic concerns the prediction of elastic properties of short-fiber reinforced composites using mean-field approaches under consideration of experimentally determined microstructure data. For this purpose, the microstructure of injection-molded specimens made of polypropylene reinforced with 30wt% of short glass fibers have been analyzed by micro-computed tomography measurements. Applying a segmentation algorithm, the spatial position, the orientation distribution and the length of the fibers have been determined. This data is evaluated in terms of orientation tensors, and length distribution and is used within three mean-field approaches: a self-consistent homogenization method, the interaction direct derivative estimate based on the three-phase model, and a two-step bounding method. All methods account for the orientation, the length, and the diameter distribution. The numerical results are compared to experimental tensile tests.

In the second topic, a detailed comparison of the predictions of linear elastic properties of mean-field homogenization approaches and full-field voxel-based homogenization methods for short-fiber reinforced materials is given. In the former case, the self-consistent, the interaction direct derivative, and a two-step-bounding approach, applying the Hashin-Shtrikman bounds, are used. In the latter case, the boundary value problem for representative volume elements is solved using fast Fourier transformation. Model microstructures with unidirectionally aligned and two misaligned fiber configurations are considered exemplarily. Fiber volume fractions of 13%, 17% and 21% and phase contrasts of 44, 100 and 1000 in the elastic moduli have been taken into account. The different homogenization schemes are compared by means of effective directional dependent Young's modulus.

In the last topic, it is discussed, whether it is sufficient to consider only the second-order fiber orientation tensor as microstructure variable describing the orientation distribution of short-fiber reinforced composites in the prediction of effective elastic properties? This question is addressed in the context of SFRCs with an overall transversal symmetric orientation distribution of fibers and, hence, effective transversal isotropic properties. Using the maximum entropy principle, it is shown, how the fiber orientation distribution function can be estimated by relying on the second-order tensor and/or the fourth-order orientation tensor, only. Both estimates are used within the self-consistent and the interaction direct derivate approach to calculate the effective linear elastic properties.

Taking all results together, the most important conclusions are the following: precise predictions of the effective elastic properties by mean-field methods are possible, if the homogenization approaches account for detailed microstructure data, which, in the best case, is determined experimentally. The experimentally determined microstructure data should be appraised with regard to the orientation distribution, volume fraction and length distribution of the fibers. In the case of a description of the orientation distribution of fibers by orientation tensors, it is important to consider orientation tensors of higher order. Fourth-order tensors capture relevant microstructural information, which may not be included in orientation tensors of second order. The combination of different fourth-order tensors with the same second-order tensor yields fundamentally different orientation distributions of fibers. This leads to significantly varying mechanical properties, and it reveals a wide range of opportunities for material design of fiber reinforced composites. For realistic material properties, the applied mean-field and full-field methods give comparable results.

The key achievement of this work is the qualitative and quantitative comparison of mean-field, full-field and experimental methods for the case of SFRC by combining well established methods from mechanics, applied mathematics and materials science.

Bibliography

Advani, S. G., Tucker III, C. L., 1987. The use of tensors to describe and predict fiber orientation in short fiber composites. Journal of Rheology 31 (8), 751–784.

Advani, S. G., Tucker III, C. L., 1990. Closure approximations for three-dimensional structure tensors. Journal of Rheology 34 (3), 367–386.

Agoras, M., Ponte Castañeda, P., 2011. Homogenization estimates for multi-scale nonlinear composites. European Journal of Mechanics - A/Solids 30 (6), 828–843.

Andrä, H., Combaret, N., Dvorkin, J., Glatt, E., Han, J., Keehm, Y., Krzikalla, F., Lee, M., Madonna, C., Marsh, M., Tapan Mukerji, Saenger, E. H., Sain, R., Saxena, N., Ricker, S., Wiegmann, A., Zhan, X., 2013a. Digital rock physics benchmarks - Part II: Computing effective properties. Computers & Geosciences 50, 33 – 43.

Andrä, H., Grzhibovskis, R., Rjasanow, S., 2013b. Boundary element method for linear elasticity with conservative body forces. In: Apel, T., Steinbach, O. (Eds.), Advanced Finite Element Methods and Applications. Vol. 66 of Lecture Notes in Applied and Computational Mechanics. Springer Berlin Heidelberg, pp. 275–297.

Bakhvalov, N., Panasenko, G., 1989. Homogenisation: Averaging processes in periodic media: Mathematical problems in the mechanics of composite materials. Mathematics and its applications. Soviet series. Kluwer Academic Publishers.

Bay, R. S., Tucker III, C. L., 1992. Stereological measurement and error estimates for three-dimensional fiber orientation. Polymer Engineering & Science 32 (4), 240–253.

Becker, F., 2009. Entwicklung einer Beschreibungsmethodik für das mechanische Verhalten unverstärkter Thermoplaste bei hohen Deformationsgeschwindigkeiten. PhD-Thesis, Martin-Luther-Universität Halle-Wittenberg, Zentrum für Ingenieurwissenschaften.

Bernasconi, A., Cosmi, F., Dreossi, D., 2008. Local anisotropy analysis of injection moulded fibre reinforced polymer composites. Composites Science and Technology 68 (12), 2574–2581.

Bernasconi, A., Cosmi, F., Hine, P., 2012. Analysis of fibre orientation distribution in short fibre reinforced polymers: A comparison between optical and tomographic methods. Composites Science and Technology 72 (16), 2002–2008.

Bishop, J. F. W., Hill, R., 1951a. A theoretical derivation of the plastic properties of a polycrystalline face-centred metal. Philos. Mag. (7) 42, 1298–1307.

Bishop, J. F. W., Hill, R., 1951b. A theory of the plastic distortion of a polycrystalline aggregate under combined stresses. Philos. Mag. (7) 42, 414–427.

Brisard, S., Dormieux, L., 2010. FFT-based methods for the mechanics of composites: A general variational framework. Computational Materials Science 49 (3), 663–671.

Budiansky, B., 1970. Thermal and Thermoelastic Properties of Isotropic Composites. Journal of Composite Materials 4 (3), 286–295.

Bull, D. J., Helfen, L., Sinclair, I., Spearing, S. M., Baumbach, T., 2013. A comparison of multi-scale 3d X-ray tomographic inspection techniques for assessing carbon fibre composite impact damage. Composites Science and Technology 75, 55–61.

Bull, D. J., Spearing, S. M., Sinclair, I., 2014. Observations of damage development from compression-after-impact experiments using ex situ micro-focus computed tomography. Composites Science and Technology 97, 106–114.

Böhlke, T., 2005. Application of the maximum entropy method in texture analysis. Computational Materials Science 32 (3–4), 276–283.

Böhlke, T., 2006. Texture simulation based on tensorial Fourier coefficients. Computers & Structures 84 (17–18), 1086–1094.

Böhlke, T., Brüggemann, C., 2001. Graphical representation of the generalized Hooke's law. Technische Mechanik 21 (2), 145–158.

Böhlke, T., Lobos, M., 2014. Representation of Hashin–Shtrikman bounds of cubic crystal aggregates in terms of texture coefficients with application in materials design. Acta Materialia 67, 324–334.

Christensen, R. M., Lo, K. H., 1979. Solutions for effective shear properties in three phase sphere and cylinder models. Journal of the Mechanics and Physics of Solids 27 (4), 315–330.

Chung, D. H., Kwon, T. H., 2002. Invariant-based optimal fitting closure approximation for the numerical prediction of flow-induced fiber orientation. Journal of Rheology 46 (1), 169–194.

Cintra, J. S., Tucker III, C. L., 1995. Orthotropic closure approximations for flow-induced fiber orientation. Journal of Rheology 39 (6), 1095–1122.

Cioranescu, D., Donato, P., 1999. An introduction to homogenization. Oxford Lecture Series in Mathematics and Its Applications. Oxford University Press.

Clarke, A., Archenhold, G., Davidson, N., 1995. A novel technique for determining the 3d spatial distribution of glass fibres in polymer composites. Composites Science and Technology 55 (1), 75–91.

Clarke, A., Davidson, N., Archenhold, G., 1993. Measurements of fibre direction in reinforced polymer composites. Journal of Microscopy 171 (1), 69–79.

Cowin, S., 1989. Properties of the anisotropic elasticity tensor. Quarterly Journal of Mechanics and Applied Mathematics 42 (2), 249–266.

De Monte, M., Moosbrugger, E., Quaresimin, M., 2010. Influence of temperature and thickness on the off-axis behaviour of short glass fibre reinforced polyamide 6.6 – cyclic loading. Composites Part A: Applied Science and Manufacturing 41 (10), 1368–1379.

Demirci, E., Acar, M., Pourdeyhimi, B., Silberschmidt, V. V., 2011. Finite element modelling of thermally bonded bicomponent fibre nonwovens: Tensile behaviour. Computational Materials Science 50 (4), 1286–1291.

DIN EN ISO 527 1-4, 2012. Plastics - Determination of tensile properties.

Dirrenberger, J., Forest, S., Jeulin, D., 2014. Towards gigantic RVE sizes for 3d stochastic fibrous networks. International Journal of Solids and Structures 51 (2), 359–376.

Doghri, I., Friebel, C., 2005. Effective elasto-plastic properties of inclusion-reinforced composites. Study of shape, orientation and cyclic response. Mechanics of Materials 37 (1), 45–68.

Dray, D., Gilormini, P., Régnier, G., 2007. Comparison of several closure approximations for evaluating the thermoelastic properties of an injection molded short-fiber composite. Composites Science and Technology 67 (7–8), 1601–1610.

Du, D.-X., Zheng, Q.-S., 2002. A further exploration of the interaction direct derivative (IDD) estimate for the effective properties of multiphase composites taking into account inclusion distribution. Acta Mechanica 157 (1-4), 61–80.

Duschlbauer, D., Böhm, H. J., Pettermann, H. E., 2006. Computational Simulation of Composites Reinforced by Planar Random Fibers: Homogenization and Localization by Unit Cell and Mean Field Approaches. Journal of Composite Materials 40 (24), 2217–2234.

Dvorak, G., Dec. 2012. Micromechanics of Composite Materials. Springer Science & Business Media.

Düster, A., Sehlhorst, H.-G., Rank, E., 2012. Numerical homogenization of heterogeneous and cellular materials utilizing the finite cell method. Computational Mechanics 50 (4), 413–431.

Eberhardt, C., Clarke, A., Vincent, M., Giroud, T., Flouret, S., 2001. Fibre-orientation measurements in short-glass-fibre composites-II: a quantitative error estimate of the 2d image analysis technique. Composites Science and Technology 61 (13), 1961–1974.

Eshelby, J. D., 1957. The Determination of the Elastic Field of an Ellipsoidal Inclusion, and Related Problems. Proceedings of the Royal Society of London. Series A. Mathematical and Physical Sciences 241 (1226), 376 –396.

Eyre, D. J., Milton, G. W., 1999. A fast numerical scheme for computing the response of composites using grid refinement. The European Physical Journal - Applied Physics 6 (01), 41–47.

Fakirov, S., Fakirova, C., 1985. Direct determination of the orientation of short glass fibers in an injection-molded poly(ethylene terephthalate) system. Polymer Composites 6 (1), 41–46.

Frieden, B., 1972. Restoring with Maximum Likelihood and Maximum Entropy. Journal of the Optical Society of America 62 (4), 511–518.

Frigo, M., Johnson, S., 1998. FFTW: An adaptive software architecture for the FFT. In: Acoustics, Speech and Signal Processing, 1998. Proceedings of the 1998 IEEE International Conference on Acoustics, Speech, and Signal Processing. Vol. 3. pp. 1381–1384 vol.3.

Frigo, M., Johnson, S., 2005. The design and implementation of FFTW3. Proceedings of the IEEE 93 (2), 216–231.

Fu, S.-Y., Lauke, B., 1996. Effects of fiber length and fiber orientation distributions on the tensile strength of short-fiber-reinforced polymers. Composites Science and Technology 56 (10), 1179–1190.

Fu, S.-Y., Lauke, B., Mäder, E., Yue, C.-Y., Hu, X., 2000. Tensile properties of short-glass-fiber- and short-carbon-fiber-reinforced polypropylene composites. Composites Part A: Applied Science and Manufacturing 31 (10), 1117–1125.

Geers, M. G. D., Kouznetsova, V. G., Brekelmans, W. A. M., 2010. Multi-scale computational homogenization: Trends and challenges. Journal of Computational and Applied Mathematics 234 (7), 2175–2182.

Ghosh, S., Lee, K., Moorthy, S., 1995. Multiple scale analysis of heterogeneous elastic structures using homogenization theory and Voronoi cell finite element method. International Journal of Solids and Structures 32 (1), 27 – 62.

Ghossein, E., Lévesque, M., 2012. A fully automated numerical tool for a comprehensive validation of homogenization models and its application to spherical particles reinforced composites. International Journal of Solids and Structures 49 (11–12), 1387–1398.

Ghossein, E., Lévesque, M., 2014. A comprehensive validation of analytical homogenization models: The case of ellipsoidal particles reinforced composites. Mechanics of Materials 75, 135–150.

Gillman, A., Matouš, K., Atkinson, S., 2013. Microstructure-statistics-property relations of anisotropic polydisperse particulate composites using tomography. Physical Review E 87 (2), 022208.

Glöckner, R., Heiliger, C., Kolling, S., 2013. A Monte-Carlo algorithm for 3d fibre detection from microcomputer-tomography. submitted.

Gross, D., Seelig, T., 2007. Bruchmechanik : mit einer Einführung in die Mikromechanik.

Guedes, J. M., Kikuchi, N., 1990. Preprocessing and postprocessing for materials based on the homogenization method with adaptive finite element methods. Computer Methods in Applied Mechanics and Engineering 83 (2), 143 – 198.

Gurp, M. v., 1998. Letter to the Editor: On the use of spherical tensors and the maximum entropy method to obtain closure for anisotropic liquids. Journal of Rheology (1978-present) 42 (5), 1269–1271.

Hahn, T., Jun. 2005. Cuba - a library for multidimensional numerical integration. Computer Physics Communications 168 (2), 78–95.

Hain, M., Wriggers, P., 2008. Computational homogenization of micro-structural damage due to frost in hardened cement paste. Finite Elements in Analysis and Design 44 (5), 233–244.

Hashin, Z., Shtrikman, S., 1962a. On some variational principles in anisotropic and nonhomogeneous elasticity. Journal of the Mechanics and Physics of Solids 10 (4), 335–342.

Hashin, Z., Shtrikman, S., 1962b. A variational approach to the theory of the elastic behaviour of polycrystals. Journal of the Mechanics and Physics of Solids 10 (4), 343–352.

Hashin, Z., Shtrikman, S., 1963. A variational approach to the theory of the elastic behaviour of multiphase materials. Journal of the Mechanics and Physics of Solids 11 (2), 127–140.

Helfen, C. E., Diebels, S., 2014. Computational homogenisation of composite plates: Consideration of the thickness change with a modified projection strategy. Computers & Mathematics with Applications 67 (5), 1116–1129.

Hill, R., 1963. New derivations of some elastic extremum principles. Progress in applied mechanics : the Prager anniversary volume. Macmillan, New York., pp. 99–106.

Hill, R., 1965. A self-consistent mechanics of composite materials. Journal of the Mechanics and Physics of Solids 13 (4), 213 – 222.

Hine, P. J., Lusti, H. R., Gusev, A. A., 2004. On the possibility of reduced variable predictions for the thermoelastic properties of short fibre composites. Composites Science and Technology 64 (7–8), 1081–1088.

Hori, M., Nemat-Nasser, S., 1999. On two micromechanics theories for determining micro–macro relations in heterogeneous solids. Mechanics of Materials 31 (10), 667–682.

Horst, J. J., 1997. Influence of fibre orientation on fatigue of short glassfibre reinforced Polyamide. Ph.D. thesis, Delft University of Technology, Delft.

Jaynes, E. T., May 1957a. Information Theory and Statistical Mechanics. Physical Review 106 (4), 620–630.

Jaynes, E. T., Oct. 1957b. Information Theory and Statistical Mechanics. II. Physical Review 108 (2), 171–190.

Johnson, S., Frigo, M., 2007. A modified split-radix FFT with fewer arithmetic operations. Signal Processing, IEEE Transactions on 55 (1), 111–119.

Joshi, M., Maiti, S. N., Misra, A., Mittal, R. K., 1994. Influence of fiber length, fiber orientation, and interfacial adhesion on poly (butylene terephthalate)/polyethylene alloys reinforced with short glass fibers. Polymer Composites 15 (5), 349–358.

Junk, M., Budday, J., Böhlke, T., 2012. On the Solvability of Maximum Entropy Moment Problems in Texture Analysis. Mathematical Models and Methods in Applied Sciences 22 (12), 1250043.

Jördens, C., Scheller, M., Wietzke, S., Romeike, D., Jansen, C., Zentgraf, T., Wiesauer, K., Reisecker, V., Koch, M., 2010. Terahertz spectroscopy to study the orientation of glass fibres in reinforced plastics. Composites Science and Technology 70 (3), 472–477.

Kabel, M., Andrä, H., 2013. Fast numerical computation of precise bounds of effective elastic moduli. Report of the Fraunhofer ITWM 224.

Kaiser, J.-M., Stommel, M., 2012. Micromechanical modeling and strength prediction of short fiber reinforced polymers. Journal of Polymer Engineering 32 (1), 43–52.

Kanatani, K.-I., 1984. Distribution of directional data and fabric tensors. International Journal of Engineering Science 22 (2), 149–164.

Kanouté, P., Boso, D. P., Chaboche, J. L., Schrefler, B. A., 2009. Multiscale Methods for Composites: A Review. Archives of Computational Methods in Engineering 16 (1), 31–75.

Kröner, E., 1971. Statistical continuum mechanics. Springer-Verlag, Wien.

Kröner, E., 1977. Bounds for Effective Elastic Moduli of Disorderd Materials. Journal of the Mechanics and Physics of Solids 25, 137–155.

Kullback, S., 1968. Information Theory and Statistics. Courier Corporation.

Laspalas, M., Crespo, C., Jimenez, M. A., Garcia, B., Pelegay, J., 2008. Application of micromechanical models for elasticity and failure to short fibre reinforced composites. Numerical implementation and experimental validation. Computers and Structures 86, 977–987.

Maire, E., Withers, P. J., 2013. Quantitative X-ray tomography. International Materials Reviews 59 (1), 1–43.

Mardia, K. V., Jupp, P. E., Sep. 2009. Directional Statistics. John Wiley & Sons.

Math2Market, 2014. Geodict.
URL `http://www.geodict.de/`

Mead, L. R., 1986. Approximate solution of Fredholm integral equations by the maximum-entropy method. Journal of Mathematical Physics 27 (12), 2903–2907.

Michel, J. C., Moulinec, H., Suquet, P., 2001. A computational scheme for linear and non-linear composites with arbitrary phase contrast. International Journal for Numerical Methods in Engineering 52 (12), 139–160.

Miled, K., Sab, K., Le Roy, R., 2011. Effective elastic properties of porous materials: Homogenization schemes vs experimental data. Mechanics Research Communications 38 (2), 131–135.

Milton, G. W., 2002. The theory of composites / Graeme W. Milton.

Mlekusch, B., Lehner, E. A., Geymayer, W., 1999. Fibre orientation in short-fibre-reinforced thermoplastics I. Contrast enhancement for image analysis. Composites Science and Technology 59 (4), 543–545.

Mori, T., Tanaka, K., 1973. Average stress in matrix and average elastic energy of materials with misfitting inclusions. Acta Metallurgica 21 (5), 571–574.

Moulinec, H., Suquet, P., 1994. A fast numerical method for computing the linear and nonlinear mechanical properties of composites. Comptes rendus de l'Académie des sciences. Série II, Mécanique, physique, chimie, astronomie 318 (11), 1417–1423.

Mura, T., 1987. Micromechanics of Defects in Solids. Second, revised edition. Mechanics of Elastic and Inelastic Solids. Martinus Nijhoff Publishers, Dordrecht.

Müller, V., Brylka, B., Dillenberger, F., Glöckner, R., Böhlke, T., Kolling, S., 2015a. Homogenization of elastic properties of short-fiber reinforced composites based on measured microstructure data. Journal of Composite Materials.

Müller, V., Böhlke, T., 2015. Prediction of effective elastic properties of fiber reinforced composites using fiber orientation tensors. submitted.

Müller, V., Böhlke, T., Kabel, M., Andrä, H., 2015b. Homogenization of linear elastic properties of short-fiber reinforced composites - A comparison of mean field and voxel-based methods. International Journal of Solids and Structures.

Nadeau, J., Ferrari, M., Nov. 2001. On optimal zeroth-order bounds with application to Hashin–Shtrikman bounds and anisotropy parameters. International Journal of Solids and Structures 38 (44–45), 7945–7965.

Nemat-Nasser, S., Hori, M., 1999. Micromechanics : overall properties of heterogeneous materials.

Nomura, S., Kawai, H., Kimura, I., Kagiyama, M., 1970. General description of orientation factors in terms of expansion of orientation distribution function in a series of spherical harmonics. Journal of Polymer Science Part A-2: Polymer Physics 8 (3), 383–400.

Ohser, J., Mücklich, F., 2000. Statistical analysis of microstructures in materials science. John Wiley, Chichester [England]; New York.

Ohser, J., Schladitz, K., 2009. 3D images of materials structures: processing and analysis. Wiley-VCH, Weinheim.

Palanivelu, S., De Pauw, S., Van Paepegem, W., Degrieck, J., Van Ackeren, J., Kakogiannis, D., Wastiels, J., Van Hemelrijck, D., Vantomme, J., 2009. Validation of digital image correlation technique for impact loading applications. Vol. 1. EDP Sciences, Brussels, Belgium, pp. 373–379.

Pierard, O., Friebel, C., Doghri, I., 2004. Mean-field homogenization of multi-phase thermo-elastic composites: a general framework and its validation. Composites Science and Technology 64 (10–11), 1587–1603.

Ponte Castañeda, P., Suquet, P., 1997. Nonlinear composites. Advances in Applied Mechanics 34, 171–302.

Ponte Castañeda, P., Willis, J. R., 1995. The effect of spatial distribution on the effective behavior of composite materials and cracked media. Journal of the Mechanics and Physics of Solids 43 (12), 1919–1951.

Reuss, A., 1929. Berechnung der Fließgrenze von Mischkristallen auf Grund der Plastizitätsbedingung für Einkristalle. ZAMM -

Journal of Applied Mathematics and Mechanics / Zeitschrift für Angewandte Mathematik und Mechanik 9 (1), 49–58.

Roland, M., Kruglova, A., Harste, N., Mücklich, F., Diebels, S., 2014. Numerical simulation of AL-SI alloys with and without a directional solidification. Image Analysis & Stereology 33 (1), 29–37.

Schladitz, K., Peters, S., Reinel-Bitzer, D., Wiegmann, A., Ohser, J., 2006. Design of acoustic trim based on geometric modeling and flow simulation for non-woven. Computational Materials Science 38 (1), 56–66.

Shannon, C., 1948. A Mathematical Theory of Communication. Bell System Technical Journal 27 (3), 379–423.

Shen, H., Nutt, S., Hull, D., 2004. Direct observation and measurement of fiber architecture in short fiber-polymer composite foam through micro-CT imaging. Composites Science and Technology 64 (13–14), 2113–2120.

Spahn, J., Andrä, H., Kabel, M., Müller, R., Linder, C., 2014. Multiscale modelling of progressive damage in elasto-plastic composite materials. Barcelona, p. submitted.

Suquet, P., 1987. Elements of Homogenization for Inelastic Solid Mechanics. Sanchez-Palencia E., Zaoui A. Homogenization Techniques in Composite Media. Springer-Verlag, Berlin, pp. 194–278.

Talbot, D. R. S., Willis, J. R., 1985. Variational Principles for Inhomogeneous Non-linear Media. IMA Journal of Applied Mathematics 35 (1), 39–54.

Talbot, D. R. S., Willis, J. R., 1992. Some simple explicit bounds for the overall behaviour of nonlinear composites. International Journal of Solids and Structures 29 (14–15), 1981–1987.

Tarigopula, V., Hopperstad, O. S., Langseth, M., Clausen, A. H., Hild, F., Lademo, O.-G., Eriksson, M., 2008. A study of large plastic deformations in dual phase steel using digital image correlation and FE analysis. Experimental Mechanics 48 (2), 181–196.

Tomasetti, E., Legras, R., Nysten, B., 1998. Quantitative approach towards the measurement of polypropylene/(ethylene-propylene) copolymer blends surface elastic properties by AFM. Nanotechnology 9 (4), 305.

Torquato, S., 2002. Random Heterogeneous Materials: Microstructure and Macroscopic Properties. Springer Science & Business Media.

Tyler, D. E., 1987. Statistical analysis for the angular central Gaussian distribution on the sphere. Biometrika 74 (3), 579–589.

Ulrych, T. J., Bishop, T. N., 1975. Maximum entropy spectral analysis and autoregressive decomposition. Reviews of Geophysics 13 (1), 183–200.

Vincent, M., Giroud, T., Clarke, A., Eberhardt, C., 2005. Description and modeling of fiber orientation in injection molding of fiber reinforced thermoplastics. Polymer 46 (17), 6719–6725.

Voigt, W., 1889. Ueber die Beziehung zwischen den beiden Elasticitätsconstanten isotroper Körper. Annalen der Physik 274 (12), 573–587.

Willis, J. R., 1977. Bounds and self-consistent estimates for the overall properties of anisotropic composites. Journal of the Mechanics and Physics of Solids 25 (3), 185–202.

Willis, J. R., 1978. Variational principles and bounds for the overall properties of composites. Continuum models of discrete systems, 185–215.

Willis, J. R., 1981. Variational and Related Methods for the Overall Properties of Composites. In: Advances in Applied Mechanics. Vol. 21. Elsevier, pp. 1–78.

Willis, J. R., 1986. Variational Estimates for the Overall Response of an Inhomogeneous Nonlinear Dielectric. In: Ericksen, J. L., Kinderlehrer, D., Kohn, R., Lions, J.-L. (Eds.), Homogenization and Effective Moduli of Materials and Media. No. 1 in The IMA Volumes in Mathematics and its Applications. Springer New York, pp. 247–263.

Wu, N., 1997. The maximum entropy method. No. 32. Springer, Berlin.

Węglewski, W., Bochenek, K., Basista, M., Schubert, T., Jehring, U., Litniewski, J., Mackiewicz, S., 2013. Comparative assessment of Young's modulus measurements of metal–ceramic composites using mechanical and non-destructive tests and micro-CT based computational modeling. Computational Materials Science 77, 19–30.

Yang, C., Huang, H.-X., Li, K., 2010. Investigation of fiber orientation states in injection-compression molded short-fiber-reinforced thermoplastics. Polymer Composites 31 (11), 1899–1908.

Zeller, R., Dederichs, P. H., 1973. Elastic constants of polycrystals. physica status solidi (b) 55 (2), 831–842.

Zheng, Q. S., Du, D. X., 2001. An explicit and universally applicable estimate for the effective properties of multiphase composites which accounts for inclusion distribution. Journal of the Mechanics and Physics of Solids 49 (11), 2765–2788.

Zheng, Q.-S., Zou, W.-N., 2001. Orientation Distribution Functions for Microstructures of Heterogeneous Materials (I) - Directional Distribution Functions and Irreducible Tensors. Applied Mathematics and Mechanics 22 (8), 865–884.

Schriftenreihe Kontinuumsmechanik im Maschinenbau
Karlsruher Institut für Technologie (KIT)
(ISSN 2192-693X)

Herausgeber: Prof. Dr.-Ing. Thomas Böhlke

Die Bände sind unter www.ksp.kit.edu als PDF frei verfügbar oder als Druckausgabe bestellbar.

Band 1 Felix Fritzen
Microstructural modeling and computational homogenization of the physically linear and nonlinear constitutive behavior of micro-heterogeneous materials. 2011
ISBN 978-3-86644-699-1

Band 2 Rumena Tsotsova
Texturbasierte Modellierung anisotroper Fließpotentiale. 2012
ISBN 978-3-86644-764-6

Band 3 Johannes Wippler
Micromechanical finite element simulations of crack propagation in silicon nitride. 2012
ISBN 978-3-86644-818-6

Band 4 Katja Jöchen
Homogenization of the linear and non-linear mechanical behavior of polycrystals. 2013
ISBN 978-3-86644-971-8

Band 5 Stephan Wulfinghoff
Numerically Efficient Gradient Crystal Plasticity with a Grain Boundary Yield Criterion and Dislocation-based Work-Hardening. 2014
ISBN 978-3-7315-0245-6

Band 6 Viktor Müller
Micromechanical modeling of short-fiber reinforced composites. 2016
ISBN 978-3-7315-0454-2